ADVANCED LOGIC FOR APPLICATIONS

SYNTHESE LIBRARY

MONOGRAPHS ON EPISTEMOLOGY,

LOGIC, METHODOLOGY, PHILOSOPHY OF SCIENCE,

SOCIOLOGY OF SCIENCE AND OF KNOWLEDGE,

AND ON THE MATHEMATICAL METHODS OF

SOCIAL AND BEHAVIORAL SCIENCES

VOLUME 110

RICHARD E. GRANDY
Department of Philosophy, University of North Carolina, Chapel Hill, U.S.A.

ADVANCED LOGIC FOR APPLICATIONS

D. REIDEL PUBLISHING COMPANY
DORDRECHT-HOLLAND / BOSTON-U.S.A.

Library of Congress Cataloging in Publication Data

Grandy, Richard E.
Advanced logic for applications.

(Synthese Library; v. 110)
Bibliography: p.
Includes indexes.
1. Logic, Symbolic and mathematical. I. Title.
BC135.G7 511'.3 77-3046

ISBN 90-277-0781-2

Published by D. Reidel Publishing Company,
P.O. Box 17, Dordrecht, Holland

Sold and distributed in the U.S.A., Canada, and Mexico
by D. Reidel Publishing Company, Inc.
Lincoln Building, 160 Old Derby Street, Hingham,
Mass. 02043, U.S.A.

Printed in The Netherlands

CONTENTS

PREFACE

This book is intended to be a survey of the most important results in mathematical logic for philosophers. It is a survey of results which have philosophical significance and it is intended to be accessible to philosophers. I have assumed the mathematical sophistication acquired in an introductory logic course or in reading a basic logic text. In addition to proving the most philosophically significant results in mathematical logic, I have attempted to illustrate various methods of proof. For example, the completeness of quantification theory is proved both constructively and non-constructively and relative advantages of each type of proof are discussed. Similarly, constructive and non-constructive versions of Gödel's first incompleteness theorem are given. I hope that the reader will develop facility with the methods of proof and also be caused by reflect on their differences.

I assume familiarity with quantification theory both in understanding the notations and in finding object language proofs. Strictly speaking the presentation is self-contained, but it would be very difficult for someone without background in the subject to follow the material from the beginning. This is necessary if the notes are to be accessible to readers who have had diverse backgrounds at a more elementary level. However, to make them accessible to readers with no background would require writing yet another introductory logic text. Numerous exercises have been included and many of these are integral parts of the proofs. This seems desirable since the purpose of the book is partly to provide the reader with the confidence and ability to go on to read more condensed material on his or her own. Some of the other examples are corollaries or interesting related theorems.

My intention is that the book should be useful both as a reference work and as a text for either self-teaching or classroom use. In order to preserve maximum flexibility, chapters have been kept independent of each other where possible. Figure a indicates graphically the

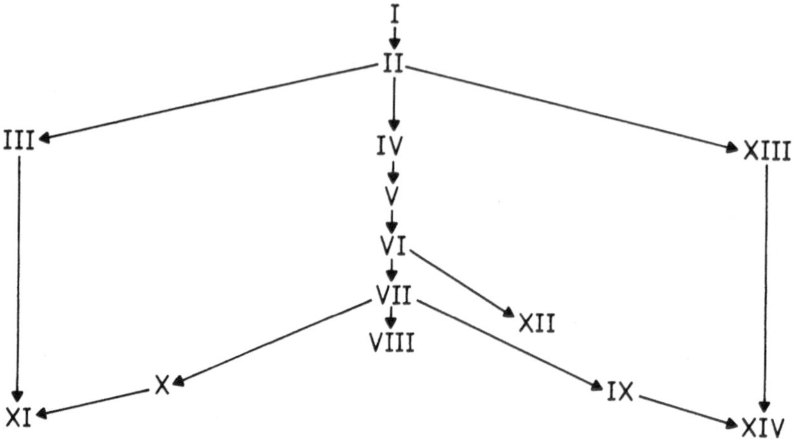

Figure a.

Material in any chapter presupposes material from those chapters which are connected to it by (a series of) arrows. For example, Chapter V presupposes I, II, and IV; Chapter XII presupposes I, II, IV, V, and VI; Chapter XIV presupposes I, II, IV, V, VI, VII, IX and also XIII.

dependencies between various chapters; I will try now to summarize the chapters and relations, and then to indicate the main types of course the book could be used for.

Chapter I isolates what I have called the Fundamental Theorem. In this theorem we characterize a particular type of set of formulas (called 'Henkin sets') and prove that these sets of formulas have interpretations. The definition of a Henkin set is entirely syntactic in the narrowest sense. That is, not only do we not mention anything about interpretations but we also make no reference to any axioms or rules. By relating the concept of a Henkin set to sets of formulas characterized in other ways, we derive the compactness and Skolem-Löwenheim theorems. Chapter I also includes a number of basic definitions required throughout the text.

Chapter II presents sets of axioms and rules of inference first for sentential calculus and then for full quantification theory and, using the Fundamental Theorem, proves the completeness of these systems. Chapter III presents an alternative formulation of first order quantification theory due to Gentzen; the completeness proof for this version of quantification theory is more closely connected with the particular rules of the system. As a consequence the proof is rather less general, but in compensation more useful corollaries concerning subsystems can be proved.

Chapters IV and V consider the extension of quantification theory to include identity and function symbols and prove some basic theorems about first order theories. The main theorems include the strong Löwenheim-Skolem theorem, the eliminability of function symbols and the partial eliminability of identity.

Chapter VI presents the general concepts and the main outline of the proof of the undecidability and incompleteness theorems. The purpose of this chapter is to give the reader overall grasp of the concepts and of the strategy of the proofs so that insight is not lost when all of the details are subsequently developed.

Chapter VII proves in detail the first Gödel theorem showing the incompleteness of any sufficiently rich number theory, Church's theorem concerning the undecidability of first order quantification theory and a number of other related theorems. Chapter VIII presents a detailed proof of Gödel's second incompleteness theorem establishing limitations on consistency proofs. Although the intuitive idea of this theorem can be stated as simply as that of the first theorem, a sufficiently accurate statement of the theorem is considerably more difficult. Considerable attention is paid to the conditions necessary for a statement to express the consistency of arithmetic. This chapter contains, to my knowledge, the first detailed textbook presentation of this theorem.

Chapter IX presents detailed proofs of Tarski's theorems, both negative and positive concerning the definability of truth. Although the topics here are somewhat independent of the previous chapters, the machinery used in proving the theorems depends heavily on previous chapters and hence cannot be read independently of them.

Chapter X extends the development of recursion theory which was begun in Chapter VII. The Kleene hierarchy is defined and various results are established concerning the undecidability of various sets and concerning the definability of recursive functions. Generalizations of Craig's theorem concerning types of axiomatizability and of Gödel's theorem are proved.

Chapter XI uses the recursive function theory just developed in order to provide a classical interpretation of intuitionistic logic and arithmetic (Kleene's recursive realizability interpretation). The independence of basic classical principles denied by intuitionists such as excluded middle and double negation is shown by means of this interpretation.

Chapter XII presents a system of second order logic, a generalization of first order logic in which quantification ranging over predicate positions is introduced. It is shown that Peano arithmetic theory based on this logic is categorical, unlike first order Peano arithmetic. It is also shown that the logic is not compact and has no recursive set of axioms. An alternative extension of first order logic which permits quantification over function symbols is also considered and shown equivalent to second order logic. Systems in which independent branches of first order quantifiers are permitted are also considered and their relation to first and second order theories is established.

Chapter XIII gives a detailed presentation of two alternative methods of formulating first order quantification in which the syntactic and semantic operations are more closely parallel. These systems are formulated in such a way that all assertions consist of equations between formulas which assert that the formulas are assigned the same truth conditions in the interpretation. In these systems the only rules of inference required are those for substitution of identities. The systems are shown to be equivalent to each other and to standard quantification theory in expressive power. In spite of their equivalence in expressive power, these systems embody a considerably different perspective on logic. From this perspective formulas are operations on sets of sequences and logic can be viewed as the study of these operations and their representation in various languages.

Chapter XIV considers a natural extension of the systems of the previous chapter which permits atomic predicates to be assigned sequences of varying length. It is shown that the standard quantification theory is properly contained in this system, that a particularly elegant definition of truth can be given in this theory, and that no recursive axiomatization of the logic exists.

Evidently this book could be used in various types of semester course emphasizing different aspects of non-elementary logic. A course in alternative forms of quantification theory could be given using Chapters I–V and Chapter XIII; a course on foundations of arithmetic could be given using Chapters I, II, IV, V–VII, X and XI; a course on first order theories could be given using Chapters I, II, IV–X; a course on alternatives to standard quantification theory could use Chapters I, II, IV–VII, X–XIV.

The Bibliographical Acknowledgements lists the original sources of the proofs and also contains indications of further material for the interested reader.

ACKNOWLEDGEMENTS

Authors whose published work I have drawn upon are listed in the Selected Bibliography, but I am also indebted to Paul Benacerraf, Alonzo Church, William Craig, Michael Dummett, Georg Kreisel, Saul Kripke and Dana Scott for lectures and conversations which have helped to shape this volume. I owe considerable gratitude to many students at Princeton University and the University of North Carolina who patiently read and improved successive approximations to the final version. The Philosophy Department of the University of North Carolina, Chapel Hill, provided financial and secretarial assistance in preparing the manuscript. John McLean compiled the indices.

CHAPTER I

HENKIN SETS AND THE FUNDAMENTAL THEOREM

We will begin by proving a fundamental result which will be used repeatedly in the proofs of our major theorems. We will prove it for the full language of quantification theory even though some of our systems will have a restricted vocabulary. No change in the proof is required for the restricted vocabularies.

The full vocabulary of quantification theory consists of the logical particles $-$, $\supset$, $\wedge$, $\vee$, $\exists$, and $\forall$, the parentheses (,), an infinite list of individual variables $x_0, x_1, x_2, \ldots$, an infinite list of individual constants $c_0, c_1, c_2, \ldots$, and for each $n > 0$ an infinite list of n-place predicate letters $F_0^n, F_1^n, \ldots$.

A *term* is any individual variable or constant.

An *atomic formula* is an n-place predicate letter followed by a sequence of n terms.

A *sentential letter* is a 0-place predicate letter.

A sequence of symbols is a *formula* iff it is atomic or if it is of the form $(A \wedge B)$ or $-A$ or $(A \supset B)$ or $(A \vee B)$ or $(\forall v)A$ or $(\exists v)A$, where A and B are formulas and v is an individual variable.

This definition illustrates our practice of using $A, B, C, D, E, A, B_1, \ldots$ as metalinguistic variables for formulas $v, v_0 \ldots$ as metalinguistic variables for object language variables. In addition we use t, $t_1, \ldots$ as metalinguistic variables for terms.

The occurrences of a variable v in a formula $(\exists v)A$ or $(\forall v)A$ are *bound occurrences*. Occurrences of a variable which are not bound are *free*. We frequently abbreviate $(\forall v)A$ by $(v)A$.

We use the notation A_t^s to indicate the formula which results from substituting t for s in A provided that if s is a variable t is substituted only for free occurrences of s, and if t is a variable all new occurrences of t are free. If these conditions are not satisfied A_t^s is simply A.

EXERCISE 1. What formula is $((x_1)\ Fx_1c_1)^{c_1}_{x_1}$?

$$((x_1)Fx_1c_1)^{c_1}_{x_2}? \qquad ((x_1)Fx_1c_1)^{x_1}_{x_2}? \qquad ((x_2)Fx_1x_2)^{x_1}_{x_2}?$$

A model for the quantificational language is an ordered pair $\langle D, I\rangle$ where D is a non-empty set and I is a function such that

(1) for each constant c, $I(c) \in D$.

(2) for each predicate letter F^n_i, $I(F^n_i) \subseteq D^n$.

D^n is the set of n-tuples of objects in D. Note that D^0 has one element, the empty sequence $\langle\ \rangle$. Thus a 0-place predicate letter F^0_i can be assigned either Λ or $\{\langle\ \rangle\}$. The first corresponds to being assigned 'true' and the second 'false' in usual presentations. The present approach may look like a 'trick' but we will show in Chapter XIII why it is natural.

In order to define truth in a model we must first define satisfaction. Let α be a function which assigns an element of D to each individual variable and $I(c)$ to each constant. Such a function is said to be a *sequence in* $\langle D, I\rangle$ and we will use the metalinguistic variables α, β, γ, α_1, β_1, ... to range over such sequences.

It will be useful to have a notation for the relation which holds between two sequences α and β when they agree on all variables except possibly v. We will write this as $\alpha \underset{v}{\approx} \beta$, and it means that for all $v' \neq v$, $\alpha(v') = \beta(v')$.

The relation *α satisfies A in $\langle D, I\rangle$* is defined recursively:

α satisfies $F^n_{t_1 \ldots t_n}$ in $\langle D, I\rangle$ iff $\langle \alpha(t_1), \ldots \alpha(t_n)\rangle \in I(F^n)$

α satisfies $-A$ in $\langle D, I\rangle$ iff α does not satisfy A in $\langle D, I\rangle$

α satisfies $(A \wedge B)$ in $\langle D, I\rangle$ iff α satisfies A in $\langle D, I\rangle$ and satisfies B in $\langle D, I\rangle$

α satisfies $(A \vee B)$ in $\langle D, I\rangle$ iff α satisfies A in $\langle D, I\rangle$ or α satisfies B in $\langle D, I\rangle$

α satisfies $(A \supset B)$ in $\langle D, I\rangle$ iff α satisfies B in $\langle D, I\rangle$ or α does not satisfy A in $\langle D, I\rangle$

α satisfies $(v)A$ in $\langle D, I\rangle$ iff for all β, if $\alpha \underset{v}{\approx} \beta$ then β satisfies A in $\langle D, I\rangle$

α satisfies $(\exists v)A$ in $\langle D, I\rangle$ iff for some β, $a \underset{v}{\approx} \beta$ and β satisfies A in $\langle D, I\rangle$.

A formula A is ***true in*** $\langle D, I\rangle$ iff A is satisfied in $\langle D, I\rangle$ by all sequences in $\langle D, I\rangle$. A formula is *valid* iff it is true in all models. We often symbolize 'A is valid' as $\models A$. A sentence is *false* in a model iff its negation is true in that model. Note that there are formulas and models such that the formula is neither true nor false in the model. We define a formula to be a *closed formula* or a *sentence* iff it has no free variables. Closed formulas are true or false in each model.

EXERCISE 2. Give an example of a formula A and model $\langle D, I\rangle$ such that A is neither true nor false in $\langle D, I\rangle$.

EXERCISE 3. Show that if A is a closed formula and $\langle D, I\rangle$ is a model then A is either true or false in $\langle D, I\rangle$.

A formula is *satisfiable* iff it is satisfied by some α in some model. A set of formulas is *simultaneously satisfiable* iff there is an α and a model such that α satisfies all of those formulas in that model. We will use Γ with and without subscripts as a metalanguage variable ranging over sets of formulas. We will use $\in$ in its usual set theoretic sense of membership.

A formula A is a semantic consequence of a set of formulas Γ iff every sequence and model that simultaneously satisfy Γ also satisfy A. This is equivalent to saying that $\Gamma \cup (-A)$ is not simultaneously satisfiable. We will often simply speak of 'consequence' meaning 'semantic consequence', and we will symbolize it as $\Gamma \models A$.

Some other conventions will be useful. We will speak of A having a model M or of M being a model for A; this means that A is satisfiable in M. Similarly we will speak of sets of formulas having a model meaning that they are simultaneously satisfied in some model. Finally we will often abbreviate simultaneously satisfiable as s.s.

Our first theorem is an intuitively plausible one which is needed frequently in our proofs. It tells us that if two formulas are alike except for their constants and free variables then if α and β assign the same things to corresponding terms then α satisfies the one formula iff β satisfies the other. More rigorously,

THEOREM. *For any model and any sequences α, β, and any formulas A, B, if*

(i) $t_1 \ldots t_n$ *do not occur in B*

(ii) $t_{n+1} \ldots t_{2n}$ *are variables which do not appear in A*

(iii) *B is* $A^{t_1 \ldots t_n}_{t_{n+1} \ldots t_{2n}}$

(iv) $\alpha(v) = \beta(v)$ *unless* $v = t_1$ *or* $\ldots$ *or* $v = t_{2n}$

(v) $\alpha(t_i) = \beta(t_{n+i})$,

then α satisfies A iff β satisfies B.

Proof. By induction on the order of formulas. An atomic formula is of order 1. If A is of order n then $-A$, $(v)A$ and $\exists vA$ are of order $n+1$. If n is the maximum of the orders of A and B then the order of $(A \wedge B)$, and $(A \supset B)$ is $n+1$.

If A is of order 1 then A is atomic and, by construction, α and β assign the same elements to the corresponding terms of A and B.

We now assume the theorem holds for orders $<n$ and show that it holds for n as well. If A is a negation, disjunction, conjunction or implication the fact to be shown follows immediately from the induction hypothesis and the definition of satisfaction. For example, if A is $-C$ then B is $-D$ where C and D satisfy the conditions of the theorem and are of order $n-1$. Therefore, α satisfies C iff β satisfies D and so α satisfies $-C$ iff β satisfies $-D$.

If A is $(v)C$ then B is $(v)D$ where C and D meet the conditions of the theorem. If $\alpha' \underset{v}{\approx} \alpha$ and $\beta' \underset{v}{\approx} \beta$ and $\alpha'(v) = \beta'(v)$ then by induction hypothesis α' sat C iff β' sat D. But then α sat $(v)C$ iff for all $\alpha' \underset{v}{\approx} \alpha$, α' satisfies C iff for all $\beta' \underset{v}{\approx} \beta$, β' satisfies D iff β satisfies $(v)D$. A similar argument establishes that α satisfies $\exists vC$ iff β satisfies $\exists vD$, if A is $\exists vC$ and B is $\exists vD$.

Intuitively, the formulas $(x_1)Fx_1x_3$ and $(x_{99})Fx_{99}x_3$ express the same thing. In general, we will say that A and B are alphabetic variants of one another if they are exactly alike except that occurrences of one or more bound variables in A are replaced by corresponding occurrences of bound variables in B. By corresponding occurrences we mean to require that all free variables of A are also free in B and that distinct bound variables of A are replaced by distinct bound variables in B.

EXERCISE 4. Which of the following are alphabetic variants of

$$(x_1)(x_3)Fx_1x_2x_3? \qquad (x_1)(x_4)Fx_1x_2x_4 \qquad (x_4)(x_3)Fx_4x_2x_3$$

$$(x_1)(x_3)Fx_1x_4x_3 \qquad (x_1)(x_1)Fx_1x_2x_1 \qquad (x_1)(x_2)Fx_1x_2x_2.$$

EXERCISE 5. Show that if A and B are alphabetic variants of one another, a sequence satisfies A iff it satisfies B.

We often want to show that a formula or set of formulas has a model. We now prove an important theorem which will be a basic tool throughout the book and we will give two applications. We are going to define a particular type of set of formulas, *Henkin sets*, and we will prove that every Henkin set has a model. Then to show that a given set of formulas has a model, it will only be necessary to show that the set in question is a subset of some Henkin set. The definition of a Henkin set is closely modeled on the definition of satisfaction in a model.

Γ is a Henkin set iff

(a) for all A either $A \in \Gamma$ or $-A \in \Gamma$
(b) for no A, $A \in \Gamma$ and $-A \in \Gamma$
(c) for all B and A, $(A \wedge B) \in \Gamma$ iff $A \in \Gamma$ and $B \in \Gamma$
(d) for all B and A, $(A \vee B) \in \Gamma$ iff $A \in \Gamma$ or $B \in \Gamma$
(e) for all B and A, $(A \supset B) \in \Gamma$ iff $A \neq \Gamma$ or $B \in \Gamma$
(f) If $A \in \Gamma$, then all formulas which are alphabetic variants of A are in Γ
(g) for all A, v, $(v)A \in \Gamma$ iff, for all terms t, $A^v_t \in \Gamma$
(h) for all A and v, $(\exists v)A \in \Gamma$ iff for some term t, $A^v_t \in \Gamma$

EXERCISE 6. Let $\langle D, I\rangle$ be a model in which every element of the domain is assigned to some constant. Show that for any α, $\{A$: α satisfies A in $\langle D, I\rangle\}$ is a Henkin set.

EXERCISE 7. Let Γ be a set of formulas such that

(1) If $(\exists v)A\Gamma$ then $A^v_c \in \Gamma$ for some c
(2) If $\Gamma \models A$ then $A \in \Gamma$
(3) For every A either A or $-A$ but not both are in Γ.
Show that Γ is a Henkin set.

We will now prove the main fact about Henkin sets.

FUNDAMENTAL THEOREM. *Every Henkin set has a model, i.e. if Γ is a Henkin set then there is an α and $\langle D, I\rangle$ such that every A in Γ is satisfied. Furthermore, D will be the natural numbers.*

Proof. We use the natural numbers as the domain. We assign each constant c_n the number $2n+1$ and α assigns each variable x_n the number $2n$. For each sentence letter we assign it $\{\langle\ \rangle\}$ iff it is in Γ. For each predicate letter F^n we assign the set of n-tuples such that F^n followed by the corresponding terms is in Γ. For example, if $F^2c_1c_1$, $F^2c_1c_3$ and $F^2x_1c_1$ are the only atomic formulas containing F^2 in Γ then I assigns $\{\langle 3, 3\rangle\ \langle 3, 7\rangle\ \langle 2, 3\rangle\}$ to F^2.

Now we will prove by induction on the order of formulas that $A \in \Gamma$ iff A is satisfied by α in $\langle D, I\rangle$.

ASSUMPTION. For all sentences A of order $<k$, $A \in \Gamma$ iff A is satisfied by α in $\langle D, I\rangle$.

We will now prove that the same holds for order k, by considering all cases.

Case a. A is atomic. If A is a sentence letter, then by the definition of $\langle D, I\rangle$ A is assigned $\{\langle\rangle\}$ iff $A \in \Gamma$. If A is of the form $F^n t_1 \ldots t_n$ then the n-tuple of objects assigned to $t_1 \ldots t_n$ was assigned to F^n iff $F^n t_1 \ldots t_n \in \Gamma$.

Case b. A is $-B$.

> $-B \in \Gamma$ iff $B \notin \Gamma$ by a, b in the definition of a Henkin set
>
> $B \notin \Gamma$ iff B is not satisfied by, induction assumption
>
> B is not satisfied iff $-B$ is satisfied.

Therefore $-B \in \Gamma$ iff $-B$ is satisfied by α in $\langle D, I\rangle$.

Case c. A is $(B \wedge C)$.

$(B \wedge C) \in \Gamma$ iff $B \in \Gamma$ and $C \in \Gamma$ (clause c in df. of Henkin set)

$B \in \Gamma$ iff B is satisfied induction assumption

$C \in \Gamma$ iff C is satisfied induction assumption

$(B \wedge C)$ is satisfied iff B is satisfied and C is satisfied.

Therefore $(B \wedge C) \in \Gamma$ iff $(B \wedge C)$ is satisfied.

EXERCISE 8. *Case d.* A is $(B \vee C)$.

EXERCISE 9. *Case e.* A is $(B \supset C)$.

Case f. By our previous Theorem.
Case g. A is $(v)B$.

(i) $(v)B \in \Gamma$ iff $B^v_t \in \Gamma$ for all terms t, by clause (g) of the df. of a Henkin set.
(ii) $B^v_t \in \Gamma$ iff B^v_t is satisfied by α in $\langle D, I\rangle$ by induction hypothesis.

If $\beta \underset{v}{\approx} \alpha$ then β satisfies B^v_t if for some t_1 $\alpha(t_1) = \beta(t)$ and α satisfies $B^v_{t_1}$. If d is any element of the domain D we know that d is assigned to some term t and we know that α satisfies B^v_t for that term by (i) and (ii). We need further, however, to consider the case of terms for which B^v_t is simply B because t is a variable which is bound in B. In those cases we know by clause f of the definition of a Henkin set that alphabetic variants B' of B are also in the Henkin set and we know that we can find a B' for which B'^v_t contains t free. By the induction assumption α satisfies B'^v_t. Thus for any d, if $\beta \underset{v}{\approx} \alpha$ and $\beta(t) = d$, β satisfies B^v_t and so $(v)B$ is satisfied by α in $\langle D, I\rangle$ if $(v)B \in \Gamma$.

In the other direction, if $(v)B$ is satisfied by α in $\langle D, I\rangle$, then for all terms B^v_t is satisfied by α and so by induction hypothesis B^v_t for all t and so $(v)B$ by the df. of a Henkin set.

EXERCISE 10. *Case h.* A is $(\exists v)B$.

We have shown that if for all formulas of order $<k$, $A \in \Gamma$ iff A is

satisfied in $\langle D, I\rangle$ then for all formulas of order k the same holds. Therefore by induction, for formulas of all orders, $A \in \Gamma$ iff A is satisfied by α in $\langle D, I\rangle$.

This completes the proof that every Henkin set is s.s. in a model whose domain is the natural numbers.

EXERCISE 11. Show that if Γ is a Henkin set then, if $\Gamma \models A$ then $A \in \Gamma$. (A set which has this property is said to be *closed under consequence*.)

EXERCISE 12. Show that if

(1) Γ is closed under consequence
(2) Γ is s.s., but if $A \notin \Gamma$, $\Gamma \cup \{A\}$ is not s.s.
(3) if $(\exists v)A \in \Gamma$ then for some t, $A_t^v \in \Gamma$

then Γ is a Henkin set.

EXERCISE 13. Suppose we are given an α and a $\langle D, I\rangle$. Let $\Gamma = \{A$: A is not satisfied by α in $\langle D, I\rangle\}$. Is Γ a Henkin set?

The first application we make of our lemma is to prove the

LÖWENHEIM-SKOLEM THEOREM. *If a set of sentences Δ is simultaneously satisfiable, then it is simultaneously satisfiable in a model whose domain is the natural numbers.*

Proof. We will construct a set Γ_w such that $\Delta \subseteq \Gamma_w$ and Γ_w is a Henkin set. To carry out our construction we need an infinite list of variables which do not occur free in Δ, so we first replace Δ by Δ^2, where $A \in \Delta$ iff the result of doubling the subscript of any variable in A occurs in Δ^2. E.g. $\exists x_2 Fx_1x_2x_3 \in \Delta$ iff $(\exists x_4)Fx_2x_4x_6 \in \Delta^2$. We note that Δ is satisfiable iff Δ^2 is. We also assume given an infinite list of all formulas $A_1, A_2, A_3, \ldots$ We now define a sequence of sets Γ_n by induction:

(a) $\Gamma_0 = \Delta^2$
(b) (i) $F_{n+1} = \Gamma_n$ if $\Gamma_n \cup \{A_{n+1}\}$ is not s.s.

(ii) $\Gamma_{n+1} = \Gamma_n \cup \{A_{n+1}\}$ if $\Gamma_n \cup \{A_{n+1}\}$ is s.s. and A_{n+1} is not of the form $-(v)B$

(iii) $\Gamma_{n+1} = \Gamma_n \cup \{A_{n+1}\} \cup \{-B_t^v\}$, if $\{A_{n+1}\} \cup \Gamma_n$ is s.s. and A_{n+1} is $-(v)B$, where t is the first variable not occurring free in $\Gamma_n \cup \{A_{n+1}\}$.

(Note that if we had used Δ instead of Δ^2 we could not be sure such a variable existed.)

We now let $\Gamma_w = \bigcup_n \Gamma_n$ and show that Γ_w has the properties required by the lemma. The set $\bigcup_n \Gamma_n$ is that set whose members are in some Γ_n or, in other words, $\Gamma_0 \cup \Gamma_1 \cup \ldots$

The first step is to show by induction that every Γ_n is s.s. For $n = 0$ this is immediate. For $n + 1$ it is trivial for clauses (bi) and (bii). So suppose $\Gamma_{n+1} = \Gamma_n \cup \{A_{n+1}\} \cup \{B_t^v\}$. We know that $\Gamma_n \cup \{-(v)B\}$ is s.s.; let α be a sequence which satisfies those formulas in interpretation M. Then there is a $\beta \approx \alpha$ such that β sat $-B$. Let γ be a sequence such that $\gamma \underset{t}{\approx} \alpha$ and $\gamma(t) = \beta(v)$. Since t does not occur in $\Gamma_n \cup \{-(v)B\}$, γ s.s. $\Gamma_n \cup \{-(v)B\} \cup \{-B_t^v\}$.

Therefore every Γ_n is s.s.

We now show that Γ_w has the desired properties.

$A \in \Gamma_w \Rightarrow -A \notin \Gamma_w$. If $A \in \Gamma_w$ and $-A \in \Gamma_w$ then for some n $\{A, -A\} \subset \Gamma_n$. Such a Γ_n would not be s.s.

$-A \in \Gamma_w \Rightarrow A \notin \Gamma_w$. Let A be the kth formula in our list and $-A$ the jth. We know that $\Gamma_{j-1} \cup \{-A\}$ is not s.s. If $k < j$, $\Gamma_{k-1} \subseteq \Gamma_{j-1}$; let α sat Γ_{j-1} in M, we know that α sat Γ_{k-1} in M and that α sat A; therefore $A \in \Gamma_k \subseteq \Gamma_w$. If $k > j$, let α sat Γ_{k-1} in M. Since $\Gamma_{j-1} \subseteq \Gamma_{k-1}$, α sat A in M and so $A \in \Gamma_k \subseteq \Gamma_w$.

$(A \supset B) \in \Gamma_w \Rightarrow A \notin \Gamma_w$ or $B \in \Gamma_w$. Suppose not; then $\{(A \supset B), A, -B\} \subset \Gamma_w$ and hence for some Γ_n, $\{A \supset B, A, -B\} \subseteq \Gamma_n$ but such a Γ_n is not s.s.

$A \notin \Gamma_w \Rightarrow (A \supset B) \in \Gamma_w$. If $A \notin \Gamma_w$, $-A \in \Gamma_w$. Suppose $(A \supset B) \notin \Gamma_w$, then for some Γ_n, $\{-A, -(A \supset B)\} \subseteq \Gamma_n$.

$B \in \Gamma_w \Rightarrow (A \supset B)\Gamma_w$. If not, then $\{B, -(A \supset B)\} \subseteq \Gamma_n$ for some n, but we know this is impossible.

$(v)B \in \Gamma_w \Rightarrow$ for all t, $B_t^v \in \Gamma_w$. Suppose not; then $(v)B \in \Gamma_w$, $B_t^v \notin \Gamma_w$ and so $-B_t^v \in \Gamma_w$ for some t. Hence for some Γ_n, $\{(v)B, -B_t^v\} \subseteq \Gamma_n$. But such a Γ_n is not s.s. and all Γ_n are s.s.

for all t, $B_t^v \in \Gamma_w \Rightarrow (v)B\Gamma_w$. Suppose not: then $-(v)B \not\subset \Gamma_w$ and so by clause (biii) for some $t - B_t^v \in \Gamma_w$. Hence for some n, $\{B_t^v, -B_t^v\} \subseteq \Gamma_n$.

EXERCISE 14. Prove the cases for $\vee$, $\wedge$, and $\exists$.

We will next prove the compactness theorem by similar means. In order to appreciate some of the significance of the theorem we need to consider some examples of sets of formulas. The set consisting of formulas $A_1, A_2, \ldots, A_n, -(A_1 \wedge A_2 \wedge \ldots \wedge A_n)$ is not s.s. but every proper subset is. Thus far nothing that we know rules out the possibility that there is an infinite set Δ which is not s.s. but each of whose finite subsets is s.s. This would mean that our derivation procedures would necessarily be defective for a derivation can contain only a finite number of premises. Thus if there were such a set Δ, it would be true that $\Delta \models B \wedge -B$ but we could not derive $B \wedge -B$ from Δ by any sound derivation procedure since $B \wedge -B$ is not a consequence of any finite subset. The compactness theorem shows that this situation cannot arise.

COMPACTNESS THEOREM. *If every finite subset of a set of sentences Δ is s.s. then Δ is s.s.*

Proof. We show that if every finite subset of Δ is s.s. then Δ^2 is a subset of a Henkin set. Thus by the *Fundamental Theorem* Δ^2 is s.s. and so Δ is also.

We will construct a Γ_w containing Δ^2 as we did in the previous theorem.

(a) $\Gamma_0 = \Delta^2$

(b) (i) $\Gamma_{n+1} = \Gamma_n$ if not all finite subsets of $\Gamma_n \cup \{A_{n+1}\}$ are s.s.

(ii) $\Gamma_{n+1} = \Gamma_n \cup \{A_{n+1}\}$ if every finite subset of $\Gamma_n \cup \{A_{n+1}\}$ is s.s. and A_{n+1} is not of the form $\exists vB$.

(iii) $\Gamma_{n+1} = \Gamma_n \cup \{A_{n+1}\} \cup \{B_t^v\}$ if every finite subset of $\Gamma_n \cup \{A_{n+1}\}$ is s.s. and A_{n+1} is $\exists vB$ and t is the first term not occurring in Γ_n or B.

We let $\Gamma_w = \bigcup_n \Gamma_n$. We will first show that every finite subset of every Γ_n is s.s. The only way this would fail to be true would be by an application of clause (biii) since Δ^2 is assumed to have only s.s. finite subsets and (bi) and (bii) preserved that property. We will show that if $\Gamma_n \cup \{\exists vB\} \cup \{B_t^v\}$ has a finite subset which is not s.s. then so does $\Gamma_n \cup \{\exists vB\}$ and thus clause (bii) preserves the finite s.s. property.

If there is a finite subset which is not s.s. then if θ is the part of the subset which is not s.s. with $\exists\ vB$ and B_t^v, then $\theta, \exists\ vB \models -B_t^v$. If θ and $\exists vB$ are s.s. by α in M, then α satisfies $-B_t^v$ in M. Since t does not occur in θ or $\exists\ vB$, for any $\beta \underset{t}{\approx} \theta$, β satisfies B_t^v and thus it satisfies B and so θ, $(\exists v)B \models (v) - B$. Thus $\{\theta, \exists vB\}$ could not be s.s. since θ, $\exists\ vB \models -(v) - B$ also.

Since every finite subset of Γ_w is a finite subset of some Γ_n, we also know that every finite subset of Γ_w is s.s. We now can show that Γ_w is a Henkin set.

(a) Suppose neither $A \in \Gamma_w$ nor $-A \in \Gamma_w$. Since both A and $-A$ appeared on the list both must have been omitted according to clause (bi) of the df. of Γ_{n+1}. Thus there is a Γ_i such that a finite subset of $\Gamma_i \cup \{A\}$ is not s.s. and a Γ_j such that a finite subset of $\Gamma_j \cup \{-A\}$ is not s.s. We know that either $\Gamma_i \subseteq \Gamma_j$ or conversely. Assume $\Gamma_i \subseteq \Gamma_j$. Let Δ_1 be the members of Γ_i not s.s. with A. Let Δ_2 be the members of Γ_j not s.s. with $-A$.

Then $\Delta_1 \cup \Delta_2$ is a finite subset of Γ_i which is not s.s. with either A or $-A$. But if any α s.s. $\Delta_1 \cup \Delta_2$ in a model it must satisfy either A or $-A$. Therefore $\Delta_1 \cup \Delta_2$ is not s.s. This contradicts the proof that every finite subset of every Γ_n is s.s. Thus Γ_w meets condition (a).

(b) Not both $A \in \Gamma_w$ and $-A \in \Gamma_w$. If both were in Γ_w then it would have a finite subset which is not s.s.

(c) Suppose $(A \wedge B) \in \Gamma_w$. If $A \notin \Gamma_w$ then by (a) above $-A \in \Gamma_w$, but this is impossible because then $\{(A \wedge B), -A\} \subseteq \Gamma_w$. Similarly, we know that $B \in \Gamma_w$.

Suppose A, B are both in Γ_w. Then if $(A \wedge B) \notin \Gamma_w$, $-(A \wedge B) \in \Gamma_w$ by (a). This would mean that $\{A, B, -(A \wedge B)\} \subset \Gamma_w$ but this is impossible since every finite subset of Γ_w is s.s.

(d) $(A \vee B) \in \Gamma_w$ iff $A \in \Gamma_w$ or $B \in \Gamma_w$.

Proof. EXERCISE 15.

(e) $(A \supset B) \in \Gamma_w$ iff $A \notin \Gamma_w$ or $B \in \Gamma_w$.

Proof. EXERCISE 16.

(f) $(v)A \in \Gamma_w$ iff for all t, $A_t^v \in \Gamma_w$.

Proof. EXERCISE 14. (Hint: Use the results of (g).)

(g) $\exists vA \in \Gamma_w$ iff for some t $A_t^v \in \Gamma_w$.

If $\exists vA \in \Gamma_w$ it was added at some stage Γ_n and thus A_t^v was also added.

If $A_t^v \in \Gamma_w$ and $\exists vA \in \Gamma_w$ then by (a) $-(\exists v)A \in \Gamma_w$ and Γ_w would have a finite subset $\{A_t^v, -\exists vA\}$ which is not s.s.

This completes the proof that Γ_w is a Henkin set. Thus we may apply the *Fundamental Theorem* to show that Γ_w is s.s. Thus Δ^2 is also s.s. and so Δ is.

CHAPTER II

DERIVATION RULES AND COMPLETENESS

Quantification theory can be formulated in various ways and the proofs of metatheorems can be given in different ways. In order to see the essential features and to understand the reasons for alternative approaches we will consider several.

For the sake of simplicity we will first run through the variations in the case of sentential calculus. Our first system is usually called an axiomatic or Hilbert type of system. The primitive vocabulary consists of an infinite list of sentential constants $F_1^0, F_2^0, \ldots$ the negation symbol $-$ and $\supset$.

The formulas of this language are all those formulas which can be formed from the vocabulary according to the rules in chapter I. In this language SC, every formula is also a sentence. The axioms of the language are all sentences of the following forms:

(A1) $A \supset (B \supset A)$

(A2) $(A \supset (B \supset C)) \supset ((A \supset B) \supset (A \supset C))$

(A3) $(-B \supset -A) \supset (A \supset B)$.

The only rule of inference is modus ponens:

From A and $A \supset B$ you may infer B.

We will call this system HSC for Hilbert Sentential Calculus.

Formula B is derivable from formulas $A_1, \ldots, A_n$ (called assumptions) iff there is a finite sequence $C_1, \ldots, C_m$ such that C_m is B and every C_i is an axiom, one of the A_j or follows from the previous lines in the derivation by modus ponens. We abbreviate 'B is derivable from $A_1, \ldots A_n$' by '$A_1 \ldots A_n \vdash B$'. If B is derivable from the empty set of assumptions ($\vdash B$), we say that *B is a theorem*. It is a fact, which should be familiar, that if $A_1, \ldots A_n \vdash B$ then $A_1, \ldots A_{n-1} \vdash A_n \supset B$. (The deduction theorem.)

EXERCISE 1. Show that $A \supset A$.

EXERCISE 2. Prove $(A \supset B) \supset ((B \supset C) \supset (A \supset C))$.

EXERCISE 3. Prove the deduction theorem.

EXERCISE 4. Prove $(A \supset B) \supset ((B \supset C) \supset (A \supset C))$ using the deduction theorem.

EXERCISE 5. Prove $A \supset --A$.

EXERCISE 6. Prove $--A \supset A$.

EXERCISE 7. Prove $A \supset (-A \supset B)$.

EXERCISE 8. Prove $-A \supset (A \supset B)$.

EXERCISE 9. Prove $-(A \supset B) \supset A$.

EXERCISE 10. Prove $-(A \supset B) \supset -B$.

Our first major metatheorem about HSC will be that it is sound. Since we are restricting ourselves to the vocabulary of sentential calculus we do not need most of the complexities of model structures. Thus we will only work with the portion we need. We will define a *truth assignment for SC* to be a function τ from sentences of SC onto T and F, such that $\tau(-A) = T$ iff $\tau(A) = F$.

$$\tau(A \supset B) = T \text{ iff } \tau(A) = F \text{ or } \tau(B) = T.$$

EXERCISE 11. Let $\langle D, I\rangle$ be given and let $\tau(A) = T$ iff A is true in $\langle D, I\rangle$. Show that τ is a truth assignment.

EXERCISE 12. Show that for an SC sentence A, $\models A$ iff $\tau(A) = T$ for all τ.

SOUNDNESS THEOREM FOR HSC. *If* $\vdash_{HSC} A$ *then* $\models A$.

Proof. By Exercise 12 we need only show that $\tau(A) = T$ for all τ.

We will prove this by induction on the length of derivations.

If a derivation is of length 1, then the line is an axiom. Thus we have three cases. If Axiom schema 1 was used then we must show $\tau((A \supset (B \supset A)) = T$, $\tau(A \supset (B \supset A)) = T$ iff $\tau(A) = F$ or $\tau(B \supset A) = T$ iff $\tau(A) = F$ or $\tau(B) = F$ or $\tau(A) = T$. This last clause must be true since $\tau(A)$ must be either T or F.

Axiom schema 2. $\tau(A \supset (B \supset C) \supset ((A \supset B) \supset (A \supset C)) = T$ iff $\tau(A \supset (B \supset C)) = F$ or $\tau((A \supset B) \supset (A \supset C)) = T$ iff $(\tau(A) = T$ and $\tau(B \supset C) = F)$ or $\tau(A \supset B) = F$ or $\tau(A \supset C) = T$ iff $(\tau(A) = T$ and $\tau(B) = T$ and $\tau(C) = F)$ or $(\tau(A) = T$ and $\tau(B) = F)$ or $\tau(A) = F$ or $\tau(C) = T$. One of these must hold so τ always assigns T to instances of Axiom schema 2.

EXERCISE 13. *Axiom schema 3.*

We now assume soundness for proofs of length n and prove that proofs of length $n + 1$ are also sound. A proof of length $n + 1$ can have as last line either an instance of an axiom or the result of a modus ponens inference. We have shown that all axioms are valid so we need only show that the result of MP inference is valid if its premises are.

Suppose $\tau(A) = T$ and $\tau(A \supset B) = T$. Then $\tau(A) = T$ and $(\tau(A) = F$ or $\tau(B) = T)$, and so $\tau(B) = T$.

This completes our proof of soundness and we now want to prove completeness.

COMPLETENESS THEOREM FOR HSC. *If* $\models A$ *then* $\vdash_{HSC} A$ *if* A *is an SC sentence.*

Proof. 'If $\models A$ then $\vdash A$' is equivalent to (I). 'For some τ, $\tau(A) = F$ or $\vdash A$' which is equivalent to (II). '$\vdash A$ or for some τ, $\tau(-A) = T$'. A sentence A is HSC-inconsistent if for some B, $A \vdash B$ and $A \vdash -B$. A sentence is HSC consistent if it is not inconsistent. We can show that $\vdash A$ iff $-A$ is HSC inconsistent. If $\vdash A$ then trivially $-A \vdash -A$ and $-A \vdash A$. If $-A \vdash B$ and $-A \vdash -B$ then $\vdash -A \supset B$ and $\vdash -A \supset -B$ so $\vdash -B \supset A$ and $\vdash B \supset A$ by Exercises 2, 5, and 6. And so by Exercise 8 $\vdash A$.

Thus the statement of completeness II above is equivalent to

(III) Either $-A$ is inconsistent or for some τ, $\tau(-A) = T$. Finally, III is equivalent to

(IV) If $-A$ is HSC consistent then for some τ, $\tau(-A) = T$. We will prove (IV) by showing that if $-A$ is HSC consistent then there is a Henkin set Γ_w such that $-A \in \Gamma_w$. From this it follows by Exercise 11 and the Fundamental Theorem that there is a τ such that $\tau(-A) = T$.

We assume $-A$ is HSC consistent and that we can construct an enumeration of all HSC sentences $B_1, \ldots$

(a) $\Gamma_0 = \{-A\}$

(b) $\Gamma_{n+1} = \Gamma_n \cup \{B_{n+1}\}$ if $\Gamma_n \cup \{B_{n+1}\}$ is HSC-consistent

(c) $\Gamma_{n+1} = \Gamma_n$ otherwise.

Every Γ_n is HSC consistent. If we let $\Gamma_w = \bigcup_{n=1,\ldots} \Gamma_n$ then Γ_w is also HSC consistent since any derivation of an inconsistency could use only finitely many premises and every finite subset of Γ_w is contained in some Γ_n.

To prove that Γ_w is a Henkin set we need to show that it satisfies clauses *a*, *b*, and *e*, since the others are vacuously true.

(a) Either $B \in \Gamma_w$ or $-B \in \Gamma_w$.

Suppose not. Then for some i $\Gamma_i \cup \{B\}$ is inconsistent and similarly for some j $\Gamma_j \cup \{-B\}$ is inconsistent. Either $i > j$ or $j > i$, suppose the first. Then $\Gamma_j \subseteq \Gamma_i$ and so $\Gamma_i \cup \{-B\}$ is inconsistent. Thus Γ_i, $-B \vdash C$ and Γ_i, $-B \vdash -C$ so by deduction theorem and Exercises 2, 5, 6 and 8 $\Gamma_i \vdash B$. But we also know that Γ_i, $B \vdash C$ and Γ, $B \vdash -C$ so by a similar argument $\Gamma_i \vdash -B$. Thus if Γ_i is itself consistent either B or $-B$ must be consistent with it.

(b) Not both $B \in \Gamma_w$ and $-B \in \Gamma_w$. Trivial, since Γ_w is HSC-consistent.

(c) $(A \supset B) \in \Gamma_w$ iff $A \notin \Gamma_w$ or $B \in \Gamma_w$.

EXERCISE 14. (Use Exercises 8–10.)

This finishes the proof that Γ_w is a Henkin set. Thus if $-A$ is HSC consistent it belongs to some Henkin set and so by the Fundamental Theorem and Exercise 11 that $\tau(-A) = T$ for some τ. This establishes (IV) and the completeness of HSC.

Notice that we now have a method for deciding if $\vdash A$ namely we use the truth table method to decide whether $\vdash A$ since by soundness and completeness we know that $\vdash A$ iff $\models A$. However, this procedure does not produce a proof of A. Later we will consider a different kind of completeness proof that gives a method of finding a proof.

We now proceed to a formulation of quantification theory which is an extension of HSC. The vocabulary will be that of quantification theory with $\wedge$, $\vee$ and $\exists$ omitted.

The formulation of quantification theory we will study first is an extension of HSC. The primitive vocabulary consists of $\forall$, $\supset$, $-$, (,), an infinite list of individual variables $x_1x_2x_3\ldots$, an infinite list of individual constants $c_1, c_2, \ldots$, and for each n an infinite list of n-place predicate $n > 0$ variables F_0^n, $F_1^n \ldots$

The axioms of the system are all formulas of the following form:

(A1) $A \supset (B \supset A)$
(A2) $(A \supset (B \supset C)) \supset ((A \supset B) \supset (A \supset C))$
(A3) $(-B \supset -A) \supset (A \supset B)$
(A4) $\forall vA \supset A_t^v$
(A5) $\forall v(A \supset B) \supset (A \supset (\forall v)B)$, if v is not free in A.

There are two rules of inference:

Modus Ponens $\dfrac{A, A \supset B}{B}$ Generalization $\dfrac{A}{\forall vA}$.

A *formula A is derivable from assumptions* $B_1, \ldots B_n$ (Abbreviated $B_1, \ldots B_n \vdash A$) iff there is a *finite* sequence of lines ending with A, each of which is an axiom or an assumption or is derivable from previous lines by modus ponens or generalization, (with the stipulation that the generalized variable in generalization inferences must not occur free in the assumptions). *A is a theorem* ($\vdash A$) iff it is derivable from the empty set of assumptions.

EXERCISE 15. (The deduction theorem) Prove, using only the facts given thus far, that if $B_1, \ldots B_n \vdash A$ then $B_1 \ldots B_{n-1} \vdash B_n \supset A$.

EXERCISE 16. Prove that this system HPC is sound.

EXERCISE 17. Show that HPC would not be sound if we did not place the restrictions on v in axiom schema 5.

We will prove the completeness of HPC by an extension of the method used for HSC. We will actually prove a lemma which is slightly stronger than we need for completeness in order to be able to prove some other corollaries as well. For completeness we need only show that if $-A$ is HPC consistent then it belongs to a Henkin set but we will prove instead that any set of HPC consistent formulas is contained in some Henkin set.

We will need one fact first.

THEOREM. Δ^2 *is* HPC *consistent iff* Δ *is.* (See page 8)

EXERCISE 18. Prove this theorem.

Now we will assume Δ is HPC consistent and construct a Henkin set Γ_w such that $\Delta \subseteq \Gamma_w$.

(a) $\Gamma_0 = \Delta^2$.

(b) $\Gamma_{n+1} = \Gamma_n$ if $\Gamma_n \cup \{A_{n+1}\}$ is inconsistent.

(c) $\Gamma_{n+1} = \Gamma_n \cup \{A_{n+1}\}$ if that set is consistent and A_{n+1} is not $-(v)B$.

(d) $\Gamma_{n+1} = \Gamma_n \cup \{-(v)B\} \cup \{-B_t^v\}$ otherwise where $-(v)B$ is A_{n+1} and t is not free in $\Gamma_n \cup \{A_{n+1}\}$.

We show by induction that every Γ_n is consistent.

$n = 0$: Trivial

$n + 1$: Trivial also except for Γ_{n+1}

which are formed by clause *d*. So let us suppose $\Gamma_n \cup \{-(v)B\} \cup \{-B_t^v\}$ is inconsistent. Then by the deduction theorem and sentential calculus $\Gamma_n, -(v)B \vdash B_t^v$. Since t is not free in Γ_n or $-(v)B$ it follows that Γ_n, $-(v)B \vdash (v)B$. Hence Γ_n is inconsistent – contrary to our induction hypothesis.

Suppose $\Gamma_w = \bigcup_n \Gamma_n$ were inconsistent. Since a proof contains only a finite number of formulas, all those used in the proof of the contradiction occur in some Γ_n. Therefore since each Γ_n is consistent Γ_w is also.

We now establish that Γ_w has the properties required:

$A \in \Gamma_w \Rightarrow -A \notin \Gamma_w$. If not, then $A \in \Gamma_w$ and $-A \in \Gamma_w$ and Γ_w would be inconsistent.

$-A \notin \Gamma_w \Rightarrow A \in \Gamma_w$. If $-A \notin \Gamma_w$ and $-A$ is the $n+1$ *st* formula in our list, $\Gamma_n \cup \{-A\}$ is inconsistent and so $\Gamma_n \vdash A$ and hence $\Gamma_w \vdash A$. Thus since Γ_w is consistent A must have been added to some $\Gamma_k \subseteq \Gamma_w$.

$(A \supset B) \in \Gamma_w \Rightarrow A \notin \Gamma_w$ or $B \in \Gamma_w$. Suppose not – by the previous property $\{(A \supset B), A, -B\} \subseteq \Gamma_w$ and Γ_w would be inconsistent.

$A \notin \Gamma_w \Rightarrow (A \supset B) \in \Gamma_w$. If not then $\{-A, -(A \supset B)\} \subseteq \Gamma_w$ and Γ_w would be inconsistent.

EXERCISE 19. $B \in \Gamma_w \Rightarrow (A \supset B) \in \Gamma_w$.

$(v)B \in \Gamma_w \Rightarrow$ for all t, $B^v_t \in \Gamma_w$. If not, $\{(v)B, -B^v_t\} \subseteq \Gamma_w$ for some t, and Γ_w would be inconsistent.

for all t, $B^v_t \in \Gamma_w$, then $(v)B \in \Gamma_w$. Suppose not; then $-(v)B \in \Gamma_w$ and so by clause (d) of our construction, $-B^v_t \in \Gamma_n \subseteq \Gamma_w$ for some n, $-B^v_t$ and Γ_w would be inconsistent.

We can now use our Fundamental Theorem to prove completeness.

COMPLETENESS THEOREM (Weak).

If $\models A$ *then* $\vdash A$.

Proof. Suppose $\models A$, then there is no α, M which satisfy $-A$. If $-A$ were consistent the Fundamental Theorem would give an α, M such that α sat $-A$ in M. Therefore for some B, $-A \vdash B$ and $-A \vdash -B$, and hence $\vdash A$.

Strong Completeness. If $B_1, \ldots B_n \models A$ then $B_1, \ldots B_n \vdash A$.

If $B_1, \ldots B_n \models A$, then $\models B_1 \supset (B_2 \supset \ldots (B_n \supset A) \ldots)$ and by weak completeness $\vdash B_1 \supset (B_2 \supset \ldots (B_n \supset A) \ldots)$ and so by modus ponens $B_1, \ldots B_n \vdash A$.

We can also give a new proof of compactness once we prove:

Strong Soundness of HPC. If $A_1, \ldots A_n \vdash_{\mathrm{HPC}} B$ then $A_1, \ldots A_n \models B$.

Proof. $A_1, \ldots A_n \vdash B$ iff $\vdash A_1 \supset (A_2, \ldots (A_n \supset B))$ (by deduction theorem) so $\models (A_1 \supset (\ldots A_n \supset B) \ldots)$ (by weak soundness) iff $A_1, \ldots A_n \models B$.

COMPACTNESS THEOREM. *If every finite subset of* Δ *is* s.s. *then* Δ *is* s.s.

Proof. If Δ were inconsistent, then for some finite subsets Δ_1, Δ_2, $\Delta_1 \vdash B$ and $\Delta_2 \vdash -B$. Therefore there would be a finite subset $\Delta_1 \cup \Delta_2$ such that $\Delta_1 \cup \Delta_2 \vdash B$ and $\Delta_1 \cup \Delta_2 \vdash -B$ and so $\Delta_1 \cup \Delta_2 \models B$ and $\Delta_1 \cup \Delta_2 \models -B$. But then $\Delta_1 \cup \Delta_2$ would not be s.s. Therefore Δ is consistent and by the Fundamental Theorem Δ is s.s.

EXERCISE 20. Using theorems and lemmas proved above (excluding the Löwenheim-Skolem Theorem) give a *short* proof of the Löwenheim-Skolem Theorem.

CHAPTER III

GENTZEN SYSTEMS AND CONSTRUCTIVE COMPLETENESS PROOFS

The method of proof just used is powerful and rather general in its applications. However, it does not give much detailed information about proofs in the system. Also, the system which we were using was specially tailored for the purpose of proving such metatheorems. This is inconvenient, however, if one is concerned with proving object language theorems or in analyzing the various connectives individually, and so in practice one introduces a number of defined expressions. Also, the system in its pure form treats only of theoremhood whereas in practice it is much easier to work with derivability from assumptions. [If you doubt this try proving $(p \supset q) \supset ((q \supset r) \supset (p \supset r))$ with and without the deduction theorem.]

This suggests that for some purposes it would be interesting to study a system which used consequence and which had more connectives. Our next system NDSC (Natural Deduction Sentential Calculus) will include symbols $\wedge$ and $\vee$ for conjunction and disjunction and $\vdash$ for consequence as well as all the symbols of HSC. The system has two axiom schemata:

$$\text{Ref}\quad \Gamma, A \vdash A \qquad \text{DN}\quad \Gamma, --\mathrm{A} \vdash \mathrm{A}$$

and the following twelve rules:

$$\vdash\supset\quad \frac{\Gamma, A \vdash B}{\Gamma \vdash A \supset B} \qquad \supset\vdash\quad \frac{\Gamma, B \vdash C \quad \Gamma \vdash A}{\Gamma, A \supset B \vdash C}$$

$$\vdash\wedge\quad \frac{\Gamma \vdash B \quad \Gamma \vdash C}{\Gamma \vdash B \wedge C} \qquad \wedge\vdash\quad \frac{\Gamma, A, B \vdash C}{\Gamma, A \wedge B \vdash C}$$

$$\vdash\vee L\quad \frac{\Gamma \vdash B}{\Gamma \vdash B \vee C} \qquad \vdash\vee R\quad \frac{\Gamma \vdash B}{\Gamma \vdash C \vee B}$$

$$\vee\vdash \quad \frac{\Gamma, A \vdash C \quad \Gamma, B \vdash C}{\Gamma, A \vee B \vdash C}$$

$$\vdash- \quad \frac{\Gamma, B \vdash A \quad \Gamma, B \vdash -A}{\Gamma \vdash -B} \qquad -\vdash \frac{\Gamma \vdash A}{\Gamma, -A \vdash B}$$

$$\text{cut} \quad \frac{\Gamma \vdash A \quad \Delta, A \vdash B}{\Gamma, \Delta \vdash B} \qquad \text{perm} \quad \frac{\Gamma, A, \Delta \vdash C}{\Gamma, \Delta, A \vdash C}$$

$$\text{thin} \quad \frac{\Gamma, A, A \vdash B}{\Gamma, A \vdash B}.$$

A string of formulas $B_1, \ldots B_n$ followed by the $\vdash$ sign followed by a formula will be called a *sequent*. For the moment we permit the left side of the sequent to be empty, but we require that the right side contain exactly one formula. Thus $\vdash A$ is a sequent, but $B \vdash$ and $B \vdash A_1, A_2$ are not.

A sequent $\Gamma \vdash A$ *is provable* iff there is a tree of sequents such that each sequent is an axiom or follows from the sequent(s) directly above it by one of the rules and $\Gamma \vdash A$ is the last line. *A is a theorem* iff $\vdash A$ is provable.

Semantic consequence is defined as before.

EXERCISE 1. Show that NDSC is sound.

EXERCISE 2. Show that if $\Gamma \models A$ then $\Gamma \vdash A$ is derivable.

This system of axioms and rules almost has some quite interesting properties. For each connective except disjunction there are two rules – one stating when a formula with the connective is derivable and the other stating a condition under which something is derivable from a formula with that connective. The only connective mentioned in the axioms is $-$. All of the rules except cut lead from sequents to more complex sequents in that all formulas in the premise(s) appear as subformulas of the inferred line. Except in the cases of $\vdash \vee L$, $\vdash \vee R$, $\vdash -$ and $- \vdash$ the *only* new symbol is the connective which is introduced by the rule. Except in the cases of cut, $\vdash \vee L$, $\vdash \vee R$, $\vdash -$ and

$-\vdash$ if a valid sequent $\Gamma \vdash A$ could have been derived in one step from $\Gamma \vdash B$ (and $\Delta \vdash C$) then $\Gamma \models B$ (and $\Delta \models C$). In other words, most of the rules are sound if used backwards, the only exceptions being those cases where a whole formula is introduced or eliminated.

Suppose we had a system in which *all* of the rules went from simple formulas to more complex ones and which were all backwards sound. Then given a sequence $\Gamma \vdash A$ we could construct its predecessors until we found that the branches of the tree all ended in sequents which are obviously valid, e.g. axioms, or that some branch ended up with a sequent which was obviously invalid. Showing that this process always gave either a proof or a counterexample would be to prove its completeness.

In order to find such a system we must look more closely at why some rules are backwards sound and others are not. To be sound backwards means that any τ which falsifies a premise of $\Gamma \vdash A$ also refutes $\Gamma \vdash A$. For example, to refute $\Gamma \vdash A \supset B$ we would try to find a τ which assigns T to all formulas in Γ, and is such that $\tau(A) = T$ and $\tau(B) = F$. Obviously such a τ also refutes $\Gamma, A \vdash B$. To refute Γ, A $\supset$ $B \vdash C$ requires a τ that assigns T to all elements of Γ and T to B or F to A, and F to C. Such a τ would also refute either $\Gamma, B \vdash C$ or $\Gamma \vdash A$. In general the attempt to make all formulas left of $\vdash$ true and right of $\vdash$ false can be reduced to simpler cases. Let us consider the problem rules.

To refute $\Gamma \vdash A \vee B$ requires assigning T to all elements of Γ and F to *both* A and B. If we were keeping a list of formulas to be falsified both A and B would be put on it, but of course we can't put both of them to the right of the same $\vdash$ sign. To refute $\Gamma \vdash -A$ is to assign T to Γ and F to $-A$, i.e. to assign T to A. But we can't add A to the left because then the premise for $\Gamma \vdash -A$ would have to be $\Gamma, A \vdash$ which isn't well-formed. To refute $\Gamma, -A \vdash B$ we would want to assign T to Γ and $-A$, and F to B, which means assigning T to Γ and F to both A and B, but again our rules don't permit sequents like $\Gamma \vdash A, B$.

With this motivation we will give a new system GSC+ with different rules. We replace $\vdash$ by $\rightarrow$, and we adopt the principle that if Γ and Δ are lists of formulas then $\Gamma \rightarrow \Delta$ is a well-formed sequent. The empty list is a list.

We have one axiom schema:

$$\Gamma, A \to A, \Delta$$

and thirteen rules:

$$\to\supset \quad \frac{\Gamma, A \to B, \Delta}{\Gamma \to A \supset B, \Delta} \qquad \supset\to \quad \frac{\Gamma, B \to \Delta \quad \Gamma \to A, \Delta}{\Gamma, A \supset B \to \Delta}$$

$$\to\wedge \quad \frac{\Gamma \to B, \Delta \quad \Gamma \to C, \Delta}{\Gamma \to B \wedge C, \Delta} \qquad \wedge\to \quad \frac{\Gamma, A, B \to \Delta}{\Gamma, A \wedge B \to \Delta}$$

$$\to\vee \quad \frac{\Gamma \to A, B, \Delta}{\Gamma \to A \vee B, \Delta} \qquad \vee\to \quad \frac{\Gamma, A \to \Delta \quad \Gamma, B \to \Delta}{\Gamma, A \vee B \to \Delta}$$

$$\to - \quad \frac{\Gamma, A \to \Delta}{\Gamma \to -A, \Delta} \qquad -\to \quad \frac{\Gamma \to A, \Delta}{\Gamma, -A \to \Delta}$$

$$\to\text{perm} \quad \frac{\Gamma \to \theta, A, \Delta}{\Gamma \to A, \theta, \Delta} \qquad \text{perm}\to \quad \frac{\Gamma, A, \theta \to \Delta}{\Gamma, \theta, A \to \Delta}$$

$$\to\text{thin} \quad \frac{\Gamma \to A, A, \Delta}{\Gamma \to A, \Delta} \qquad \text{thin}\to \quad \frac{\Gamma, A, A \to \Delta}{\Gamma, A \to \Delta}$$

$$\text{cut} \quad \frac{\Gamma \to A, \theta \quad \Gamma, A \to \theta}{\Gamma \to \theta}.$$

A sequent $\Gamma \to \Delta$ *is derivable* iff there is a tree whose bottom line is $\Gamma \to \Delta$ and each element of which is an axiom or else follows from immediately preceding lines. A formula A *is a theorem* iff $\to A$ is derivable. A sequent $\Gamma \to \Delta$ *is valid* ($\Gamma \models \Delta$) iff every τ which assigns T to all elements of Γ assigns T to at least one member of Δ.

EXERCISE 3. Show that $--A \to A$ is derivable.

EXERCISE 4. Show that if $\Gamma \to A$ is derivable then so is $\Gamma \to A \vee B$.

EXERCISE 5. Show that if $\Gamma, B \to A$ and $\Gamma, B \to -A$ are derivable then so is $\Gamma \to -B$.

EXERCISE 6. Show that GSC+ is sound.

EXERCISE 7. Show that all of the rules except cut are backwards sound.

In GSC+ (Gentzen Sentential Calculus with cut) all of the rules are of the sort we want except cut. We solve this problem by eliminating cut and working with GSC, only the first sixteen rules of GSC+. We will prove the completeness of GSC by constructing a process which, given a sequent $\Gamma \rightarrow \Delta$, either produces a proof of $\Gamma \rightarrow \Delta$ or a τ which assigns T to all elements of Γ and F to all elements of Δ. A few definitions will make it easier to define the process.

In a sequence $\Gamma, A \rightarrow B, \Delta$ the formula immediately preceding $\rightarrow$ will be called the *left principal formula* (LPF) and the formula immediately following $\rightarrow$ is the *right principal formula. A formula is atomic* iff it contains no connectives. *A list Γ is atomic* iff every member of Γ is atomic. For any sequent $\Gamma \rightarrow \Delta$ whose LPF (RPF) is not atomic, there is a single connective rule such that there is a (pair of) sequent(s) determined by $\Gamma \rightarrow \Delta$ from which $\Gamma \rightarrow \Delta$ follows by that rule. These sequents are the immediate left (right) predecessors of $\Gamma \rightarrow \Delta$.

We now define a process which produces a sequence T_n of trees for any sequence: $T_0 = \Gamma \rightarrow \Delta$.

If $T_{2n}(\Gamma \rightarrow \Delta)$ is the tree obtained at stage $2n$, then T_{2n+1} is defined as follows: For each sequent $\phi \rightarrow \theta$ which is a topmost node in T_{2n} and whose LPF is not atomic, we write above it its immediate predecessor. If ϕ is atomic we do nothing; if ϕ is not atomic but its LPF is atomic, i.e. $\phi = \phi', p$ then we write above $\phi \rightarrow \theta$ the sequent $p, \phi' \rightarrow \theta$. If the LPF of this sequent is complex we apply the rule above, if the new LPF is atomic we move it to the left and repeat our procedure.

T_{2n} is defined from T_{2n-1} by an analogous procedure switching right and left in the definition. We now establish some properties of the construction.

LEMMA 1. *For any sequent $\Gamma \rightarrow \Delta$ there is an n such that $T_n(\Gamma \rightarrow \Delta)$ contains only atomic formulas on the top nodes.*

Proof. At each two steps the topmost nodes have one or two fewer occurrences of a connective. Thus if $\Gamma \to \Delta$ has m occurrences of connectives $n \leqslant 2m$.

LEMMA 2. *If a sequent* $\phi \to \theta$ *is atomic then either it is derivable from an axiom in at most two steps or there is a* τ *which assigns* T *to all members of* ϕ *and* F *to all members of* θ.

Proof. If no letter occurs in both ϕ and θ we define $\tau(F^0) = T$ iff $F^0 \in \phi$ and $\tau(F^0) = F$ otherwise. If some letter p is in both ϕ and θ, then they have the form $\phi_1, F^0, \phi_2 \to \theta_1, F^0, \theta_2$ which is derivable thus:

$$\phi_1, \phi_2, F^0 \to F^0, \theta_1, \theta_2$$
$$\phi_1 F^0, \phi_2 \to F^0, \theta_1, \theta_2$$
$$\phi_1, F^0, \phi_2 \to \theta_1, F^0, \theta_2.$$

LEMMA 3. *Let* T *be the completed tree for* $\Gamma \to \Delta$, *if the top nodes of* T *are derivable, then* $\Gamma \to \Delta$ *is derivable.*

Proof. By examining the ways in which T is generated.

LEMMA 4. *If* T *is a tree generated by our definition for a sequent* $\Gamma \to \Delta$ *then if* τ *falsifies a node of* T τ *falsifies* $\Gamma \to \Delta$.

Proof. Again by inspecting the rules used in generating the tree.

EXERCISE 8. Show Lemma 4 in detail in case the procedure is operating on a LPF of the form $(A \supset B)$.

COMPLETENESS THEOREM FOR GSC. *For any sequent* $\Gamma \to \Delta$, *we can find a proof of* $\Gamma \to \Delta$ *if it is valid.*

Proof. Consider $T_n(\Gamma \to \Delta)$ for the value of n which terminates the procedure. By Lemma 1 n exists and T_n has only atomic formulas at the top nodes. If these are all derivable from axioms, then by Lemmas 2 and 3 $\Gamma \to \Delta$ is derivable.

If some sequent at a top node is not derivable from an axiom then by Lemma 2 there is a τ which falsifies that sequent and so by Lemma 4 τ falsifies $\Gamma \to \Delta$. But this is impossible if $\Gamma \to \Delta$ is valid.

COROLLARY 1. *In* GSC+, *if* $\Gamma \to \Delta$ *is derivable using* cut, *it is also derivable without using* cut.

COROLLARY 2. *If there is a proof of $\Gamma \to \Delta$ in* GSC+, *then there is a proof such that every formula which occurs in the proof is a subformula of a formula in $\Gamma \to \Delta$.*

[*A is a subformula of B* iff A is B or B is $-A$, $A \supset C$, $C \supset A$, $A \vee C$, $C \vee A$, $C \wedge A$, $A \wedge C$ or if A is a subformula of a subformula of B.]

COROLLARY 3. *There is a decision procedure for provability in* GSC+.

Proof. Constructing T_n is a mechanical task and supplemented by Lemma 2 will produce a proof if there is one and will terminate in a non-proof if there is none.

EXERCISE 9. Show that Corollary 2 does not hold for HSC.

As we promised earlier, the completeness proof we have just given for GSC provides more information about the nature of proofs in the system. The first proof assured us that if a formula was valid it could be proved, but the completeness proof gave no indication of how to find such a proof. The first method is not without its own advantages, of course, as we shall see shortly when we extend our systems to include quantifiers.

EXERCISE 10. Show that $\to (A \supset (B \supset C)) \supset ((A \supset B) \supset (A \supset C))$ is derivable by constructing the relevant T.

EXERCISE 11. Show that $\to (p \supset q) \vee (q \supset R)$ is derivable.

Our next project is to give a system of quantification theory which extends the ideas of the Gentzen Sentential Calculus. We would expect to add two new rules for each of the two quantifiers. To find the new rules we follow the 'search for a counterexample' strategy. If we want to refute a sequent $\Gamma \to (v)A, \Delta$, it suffices to make all of the formulas in Γ T and the formulas in $\{(v)A, \Delta\}F$.[1] To make $(v)A$ F it suffices to make $A_t^v F$ for some t. Thus we can reduce the problem of refuting $\Gamma \to (v)A, \Delta$ to that of refuting $\Gamma \to A_t^v, \Delta$. We must use some care in choosing t however. The sequent $A_t^v \to (v)A$ is not valid if v is free in A, but if we 'reduce' this sequent to $A_t^v \to A_t^v$ then we have a valid sequent. To avoid this difficulty we stipulate that t is a variable which is not free in Γ, A or Δ.

EXERCISE 12. Show that if t is free in Δ the rule is unsound.

For a refutation of $\Gamma, (v)A \to \Delta$ it is necessary to assign T to $(v)A$ and this requires at least that T is assigned to A^v_t for all t. We cannot do this directly because we require that our sequents be finite. If we put a finite number, e.g. one, instance in the antecedent we obtain the rule

$$\frac{\Gamma, A \to \Delta}{\Gamma, (v)A \to \Delta},$$

which is sound, but not backwards sound. The difficulty, of course, is that we need to ensure that all instances of $(v)A$ are T when we conduct our search procedure. The solution then is to keep $(v)A$ in the sequent as a reminder that we haven't finished with it. So our rule will be:

$$\forall \to \quad \frac{\Gamma, A^v_t, (v)A \to \Delta}{\Gamma, (v)A \to \Delta}.$$

[Note that repeated applications of this rule will enable us to go from Γ, $A^v_{t_1}, \ldots, A^v_{t_n}$, $(v)A \to \Delta$ to Γ, $(v)A \to \Delta$.]

Reasoning analogous to that for $\forall$ will show that the $\exists$ rules should be:

$$\to \exists \quad \frac{\Gamma \to (\exists\, v)A, A^v_t, \Delta}{\Gamma \to (\exists\, v)A, \Delta} \qquad \exists \to \quad \frac{\Gamma, A^v_t \to \Delta}{\Gamma, (\exists\, v)A \to \Delta}$$

where t is not free in Γ or Δ and t is a term.

EXERCISE 13. Show that $\to\exists$ and $\exists\to$ are sound forwards and backwards.

Our proof of the completeness of this new system will be like that for GSC, with one complication. In a strict syntactic sense the only subformulas of $(x_1)F^1_1x_1$ are itself and $F^1_1x_1$. However, there is also the semantic conception of subformula which is that of a formula which is relevant to the truth or falsity of $(x_1)F_1^1x_1$, and in this sense $F_1^1x_2, F_1^1x_3 \ldots$ are all subformulas. Looking at the rules you can see that it is only in the latter sense that our system has the subformula property. Thus each quantified formula has infinitely many subformulas and we have no guarantee our proof/refutation procedure will terminate.

GPC

Axiom $\Gamma, A \to A, \Delta$

$$\to\supset \quad \frac{\Gamma, A \to B, \Delta}{\Gamma \to A \supset B, \Delta} \qquad \supset\to \quad \frac{\Gamma, B \to \Delta \quad \Gamma \to A, \Delta}{\Gamma, A \supset B \to \Delta}$$

$$\to\wedge \quad \frac{\Gamma \to A, \Delta \quad \Gamma \to B, \Delta}{\Gamma \to A \wedge B, \Delta} \qquad \wedge\to \quad \frac{\Gamma, A, B \to \Delta}{\Gamma, A \wedge B \to \Delta}$$

$$\to\vee \quad \frac{\Gamma \to A, B, \Delta}{\Gamma \to A \vee B, \Delta} \qquad \vee\to \quad \frac{\Gamma, A \to \Delta \quad \Gamma, B \to \Delta}{\Gamma, A \vee B \to \Delta}$$

$$\to - \quad \frac{\Gamma, A \to \Delta}{\Gamma \to -A, \Delta} \qquad -\to \quad \frac{\Gamma \to A, \Delta}{\Gamma, -A \to \Delta}$$

$$\to\forall \quad \frac{\Gamma \to A_t^v \Delta}{\Gamma \to \forall v A, \Delta} \qquad \forall\to \quad \frac{\Gamma, A_t^v, \forall v A \to \Delta}{\Gamma, \forall v A \to \Delta}$$

(t not free in Γ, A or Δ).

$$\to\exists \quad \frac{\Gamma, \to \exists v A, A_t^v, \Delta}{\Delta \to \exists v A, \Delta} \qquad \exists\to \quad \frac{\Gamma, A_t^v \to \Delta}{\Gamma, \exists v A \to \Delta}$$

(t not free in Γ, A or Δ).

$$\to\text{perm} \quad \frac{\Gamma \to \theta, A, \Delta}{\Gamma \to A, \theta, \Delta} \qquad \text{perm}\to \quad \frac{\Gamma, A, \theta \to \Delta}{\Gamma, \theta, A \to \Delta}$$

$$\to\text{thin} \quad \frac{\Gamma \to A, A, \Delta}{\Gamma \to A, \Delta} \qquad \text{thin}\to \quad \frac{\Gamma, A, A \to \Delta}{\Gamma, A \to \Delta}.$$

Completeness of GPC. *If a sequent* $\Gamma \to \Delta$ *is valid then it is derivable.*

Proof. We will extend the tree construction we used for GSC; note that for principal formulas whose main connective is a sentential connective the immediately preceding sequent(s) is (are) unique, but for principal formulas whose main logical symbol is a quantifier an infinite number of sequents are immediate predecessors. Hence our procedure must be made more complex in order to specify a unique predecessor in a way that guarantees that we will obtain a proof or counter model. So we will associate with each node on our tree a sequence of lists L; each L will consist of a list of terms which appear free and for various quantified formulas we will have lists of the instances which have been used in the construction.

We will again define T_n by induction:

$$T_0(\Gamma \to \Delta) = \Gamma \to \Delta$$

$$L_0 = \{t\colon t \text{ occurs free in } \Gamma \cup \Delta \text{ or } t \text{ is a constant in } \Gamma \cup \Delta \text{ or } t \text{ is } c_0\}$$

If $T_{2n}(\Gamma \to \Delta)$ is the tree obtained at stage $2n$, then $T_{2n+1}(\Gamma \to \Delta)$ is the tree defined as follows:

For each sequent $\phi \to \theta$ which is a topmost node σ and whose LPF is not atomic and not quantificational, we write above it its immediate predecessor(s). The lists for these nodes are the same as L_σ. If the LPF is quantified and ϕ is $\exists vA$, i.e. the sequent is $\phi', \exists vA \to \theta$, then we write above it the new node $\phi', A^v_t \to \theta$, where t is the first constant not in the list of terms in L. $L_{\sigma'}$, σ' being the new node, consists of the result of adding the term t to the list of terms in L. If the LPF is universal, i.e. the sequent is $\phi', (v)A \to \theta$, then we consult L_σ: if there is a list of instances of $(v)A$ in L then we write above σ

$$\frac{\forall vA, \phi', A^v_t \to \theta}{\phi', A^v_t, \forall vA \to \theta}$$

where t is the first term which appears on the list of terms in L but whose instantiation in $(v)A$ is not listed. If there is no list of instances of $(v)A$, then t is to be the first term on the list of terms in L_σ. In the first case A^v_t is added to the list of instances; in the second case we add a new list to L_σ, i.e. $(v)A$: A^v_t, in order to obtain the new $L_{\sigma'}$. If for every term t on the list in L_σ A^v_t is already on the list of instances of $(v)A$, then we write above $\phi \to \theta$ the sequent $(v)A, \phi' \to \theta$ and $L_{\sigma'} = L_\sigma$.

If the LPF is atomic and ϕ is atomic nothing is placed above $\phi \to \theta$.

If the LPF is atomic but ϕ is not, we write $A, \phi' \to \theta$ above $\phi', A \to \theta$, set $L_{\sigma'} = L_\sigma$ and repeat the procedure above, until some non-atomic formula is operated on.

If $T_{2n-1}(\Gamma \to \Delta)$ is the tree obtained at stage $2n-1$, then $T_{2n}(\Gamma \to \Delta)$ is the tree defined as follows:

For each sequent which is a top node $\sigma\, \phi \to A, \theta$ with RPF A, we operate according to the following principles:

If A's main connective is sentential we write above σ the immediate predecessor and let the new $L = L_{\sigma'}$.

If A is $(v)B$, then we write above σ the sequent $\phi \to B^v_t, \theta$, where t is the first variable which does not appear on the list of terms in $L_{\sigma'}$. $L_{\sigma'}$, is the result of adding t to the list in $L_{\sigma'}$.

If A is $(\exists v)B$ then if for all terms t in the list L_σ B^v_t is in the list of instances of $(\exists v)B$, we write above σ, $\phi \to \theta, (\exists v)B$ and let $L_{\sigma'} = L_\sigma$. If there is no list of instances or if the list does not include B^v_t for all terms t in the list in L, then we write above $\phi \to (\exists v)B, \theta$

$$\frac{\phi \to B^v_t, \theta(\exists v)B}{\phi \to \exists vB, B^v_t, \theta}$$

and add B^v_t to the list of instances, where t is the first term on the list in L such that B^v_t is not on the list of instances in L, in order to obtain $L_{\sigma'}$.

If A is atomic and θ is atomic, then we do nothing.

If A is atomic and θ is not, then we write above $\phi \to A, \theta$ the sequent $\phi \to \theta, A$ and repeat the above procedure until a nonatomic formula is operated on.

We now must consider the form which our trees[2] can take. Consider a branch of a tree B and let B_n be the portion of the branch generated in T_n. There are three possibilities:

> For some n, B_n terminates, i.e. for all $k > n$ $B_k = B_n$.
>
> For some n, all B_k for $k > n$ are the result of writing permutations of the top node at B_n. This can occur if we have a sequent $\phi \to \theta$ where all formulas in ϕ are atomic or are universal formulas $(v)B$ such that all instances with terms in the relevant list have already been added

and all formulas in θ are atomic or are existential formulas $(\exists v)B$ such that all instances with terms on the list L have already been added. If B_n is the least B_n with this property we will say that B is cyclic beyond B_n.

For each B_n, there is an $m > n$ such that B_m is not the result of permuting sequents from earlier nodes.

Consider the completed tree $T(\Gamma \rightarrow \Delta)$; we will prune this tree by deleting those portions of branches such that the branch is cyclic beyond that point.

Let us call the pruned tree T''. Some branches of T'' may have nodes which are sequents that are permutations of axioms but where the procedure does not terminate because there are other formulas to be worked on. Such branches are effectively recognizable and we will also prune them at the point where a permutation of an axiom appears. We will call the twice pruned tree T'. This pruned tree T' could have one of four forms:

(a) Every branch terminates and all top nodes are of the form $\phi_1, A, \phi_2 \rightarrow \theta_1, A, \theta_2$, and T' is finite.

(b) At least one branch terminates and its top node is not of the form $\phi_1, A, \phi_2 \rightarrow \theta_1, A, \theta_2$.

(c) There is at least one infinite branch.

(d) Every branch terminates but T' is infinite.

We shall show that in case (a) $\Gamma \rightarrow \Delta$ is derivable, in cases (b) and (c) we can define an α and M such that α satisfies Γ in M and does not satisfy any element of Δ in M. We will show that case (d) is impossible.

KÖNIGS LEMMA. *An infinite binary tree has at least one infinite branch.*

Proof. Suppose we are given an infinite tree T – we will show that there is a branch B such that for each n the nth node in B has the property that there are infinitely many nodes above it. Such a branch must be infinite since on a finite branch there is at least one point with *no* nodes above it.

We define B inductively. The first node of B is the bottom of the tree – clearly this point has infinitely many nodes above it. Suppose B_n is defined – we will show how to find B_{n+1}.

If the top node of B_n has one node immediately above it, that node has infinitely many nodes above it. If the top node of B_n has two nodes above it σ_l and σ_r, consider the leftmost one σ_l. If σ_l has infinitely many nodes above it we add σ_l to B_n to form B_{n+1}. If σ_l does not have infinitely many nodes above it, then σ_r must and we form B_{n+1} by adding σ_r to B_{n+1}.

Therefore case (d) is impossible.

Cases (b) and (c). Let B be a branch which is either infinite or which terminates in a sequent which is not a permutation of an axiom. Let L be the list associated with the last node, if there is one, and let it be the union of the lists associated with B if B is infinite. What we wish to do is construct an α and M which show that the sequent $\Gamma \rightarrow \Delta$ is not valid. In order to do this we first form the set ϕ of formulas which appear on the right of a sequent in B and the set θ of formulas which appear on the left of such a sequent. We note that

(1) if $(A \vee B) \in \theta$ then $A \in \theta$ or $B \in \theta$
(2) if $(A \wedge B) \in \theta$ then $A \in \theta$ and $B \in \theta$
(3) if $(A \supset B) \in \theta$ then $B \in \theta$ or $A \in \phi$
(4) if $-A \in \theta$ then $A \in \phi$
(5) if $A \vee B \in \phi$ then $A \in \phi$ and $B \in \phi$
(6) if $A \wedge B \in \phi$ then $A \in \phi$ or $B \in \phi$
(7) if $(A \supset B) \in \phi$ then $A \in \theta$ and $B \in \phi$
(8) if $-A \in \phi$ then $A \in \theta$
(9) if $(v)A \in \theta$ then $A_t^v \in \theta$ for all t on L
(10) if $\exists vA \in \theta$ then $A_t^v \in \theta$ for some t in L
(11) if $(v)A \in \phi$ then $A_t^v \in \phi$ for some t in L
(12) if $(\exists v)A \in \phi$ then $A_t^v \in \phi$ for all t in L.

For any formula A in θ there is a sequent $\theta_1, A, \theta_2 \rightarrow \phi_1$ where θ_2 contains n connectives. In at most $2n$ steps a sequent $\theta_3, A \rightarrow \phi_2$ will be reached and by checking the respective clauses for various sentential connectives and $\exists$ one can verify 1–4 and 10. A similar argument on the right of $\rightarrow$ verifies 5–8 and 11. If we consider a formula $(v)A \in \theta$ and let t be the first term which has not been substituted for v in θ then the argument above shows that in a finite number of steps $A_t^v \in \theta$. Thus by induction each t is instantiated in some finite number of steps; consequently infinite branches contain all instances of A in θ if $(v)A \in \theta$, and finite branches do not terminate until all instances have been added. A parallel argument can be given for $(\exists v)A \in \phi$, establishing 12).

We now construct an α and M analogous to those in the theorem on page 6, except that we confine our attention to subformulas of $\Gamma \to \Delta$.

We let $I(c_n) = 2n$ if $c_n \in L$ $\quad D = \{n: I(t) = n$ for some $t\}$

$$I(c_n) = 0 \quad \text{otherwise}$$

$$\alpha(x_n) = 2n + 1 \quad \text{if } x_n \in L$$

$$\alpha(x_n) = 0 \quad \text{otherwise}$$

$$I(F^n) = \{\langle \alpha(t_1), \ldots \alpha(t_n) \rangle : Ft_1 \ldots t_n \in \theta\}.$$

We prove by induction on the number of connectives that α sat A in M if $A \in \theta$, and if $A \in \phi$ α does not satisfy A in M.

If $n = 0$, α sat $Ft_1 \ldots t_n$ iff $\langle \alpha(t_1) \ldots \alpha(t_n) \rangle \in I(F^n)$, which holds iff $Ft_1 \ldots t_n \in \theta$. If $Ft_1 \ldots t_n \in \phi$ then $Ft_1 \ldots t_n \in \theta$.

If $n = k + 1$, we proceed by cases.

Case 1. $A \vee B \in \theta$. In this case $A \in \theta$ or $B \in \theta$ and by induction hypothesis α sat A or α sat B.

Case 5. $A \vee B \in \phi$. In this case $A \in \phi$ and $B \in \phi$ and so by induction hypothesis α does not satisfy either A or B.

Case 9. $(v)A \in \theta$. By (9) in the lemma $A_t^v \in \theta$ for all t on L, and so if $\beta \underset{v}{\approx} \alpha$ and β assigns $\alpha(t)$ to v, β satisfies A. The class of such β's exhausts the β's such that $\beta \underset{v}{\approx} \alpha$, so for every $\beta \underset{v}{\approx} \alpha$, β sat A and hence α satisfies $(v)A$.

EXERCISE 14. *Case 12.*

Assuming that all cases can be proved, we have shown that with trees of type (b) or (c) we can construct an α and M such that α

satisfies all members of Γ and none of Δ and hence we know that $\Gamma \to \Delta$ is not valid.

In case (a) we have a finite tree each top node of which is of the form $\theta_1, A, \theta_2 \to \phi_1, A, \phi_2$. Obviously these sequents are derivable from axioms (by permutation where necessary) and since our construction is such that every sequent is derivable from those above it, we obtain a proof of $\Gamma \to \Delta$.

Thus since case (d) is impossible we either know that (cases b, c) $\Gamma \to \Delta$ is not valid or (case a) it is provable. Thus GPC is complete.

We have not discovered a decision procedure for validity of formulas of GPC, however. Our procedure does not always terminate in a finite number of steps. This does not establish that there is no decision procedure for we do not know at this point whether our failure to find one is due to our lack of imagination or to the nonexistence of a method.

EXERCISE 15. Give a sequent $\Gamma \to \Delta$ for which the procedure does not terminate in a finite number of steps.

There are, however, some special cases for which our tree method does give a decision procedure. For example, suppose A is of the form $(v_1)(v_2)\ldots(v_n)B$ where B contains no quantifiers. In this case $T_{2n+1}(\to A)$ is a tree whose top node is $B^{v_1 \ldots v_n}_{t_1 \ldots t_n}$; this last formula will be reduced to simpler formulas in a finite number of steps and we will obtain a finite tree. From this tree we can construct either a proof of $\to A$ or an α and M such that α fails to satisfy A in M.

THEOREM. *There is a decision procedure for validity of formulas of the form* $(v_1)\ldots(v_n)B$ *where* B *contains no quantifiers.*

THEOREM. *There is a decision procedure for validity of formulas of the form* $(\exists v_1)\ldots(\exists v_n)B$ *where* B *contains no quantifiers*

EXERCISE 16. Prove this theorem.

THEOREM. *There is a decision procedure for formulas of the form* $(v_1) \ldots (v_n)(\exists\, v_{n+1}) \ldots (\exists\, v_{n+m})B$ *where B contains no quantifiers.*

EXERCISE 17. Prove this theorem.

We have been using the fact that our proof of completeness shows that the system has the subformula property (cf. p. 29), but we can show even stronger results.
A is in *prenex normal form* iff A is of the form (Q_1v_1) $(Q_2v_2) \ldots (Q_nv_n)B$ where each Q is $\forall$ or $\exists$ and B contains no quantifiers. It can be shown that for any formula C there is a formula A in prenex normal form such that $A \equiv C$.

THEOREM. *If A is in prenex normal form and* $\rightarrow A$ *is derivable, then there is a proof which contains a sequent* $\rightarrow \Delta$ *such that the only rules used in deriving* $\rightarrow \Delta$ *are sentential rules and permutation, and the only rules used in the derivation of* $\rightarrow A$ *from* $\rightarrow \Delta$ *are quantificational permutation, and thinning.*

EXERCISE 18. Prove this theorem.

NOTE

[1] For brevity, I will talk of making ΓT when I mean making all members of ΓT.
[2] A tree is a set with a discrete partial ordering such that a) exactly one point has no predecessor b) every point has zero or a finite number of successors.

CHAPTER IV

QUANTIFICATION THEORY WITH IDENTITY AND FUNCTIONAL CONSTANTS

We wish now to extend our theory to include identity and functions. The system will be an extension of HPC which we will call HPC$^=$. The primitive symbols are $-$, $\supset$, $\forall$,), (, and $=$; an infinite list of individual variables $x_0, x_1, x_2, \ldots$, an infinite list of predicate letters F_0^n, F_1^n, $F_2^n, \ldots$ for each $n > 0$, an infinite list of constants, $c_0, c_1, \ldots$ and for each $n > 0$ an infinite list of function letters $f_0^n, f_1^n, \ldots$

We define formula and term by inductive definitions:

> Each individual variable and constant is a term; if $t_1 \ldots t_n$ are terms then $f_i^n(t_1, \ldots t_n)$ is a term.
>
> If $t_1, \ldots t_n$ are terms then $F_i^n t_1 \ldots t_n$ and $t_1 = t_2$ are formulas; if A and B are formulas then so are $-A$, $(A \supset B)$ and $(v)A$.

The axioms will be those of HPC plus the axiom $(x_1)(x_1 = x_1)$ and the schema

$$(v_1)(v_2)[v_1 = v_2 \supset (A \supset A_{v_2}^{v_1})].$$

The rules of inference remain the same. We must, however, revise the definition of A_t^s slightly: A_t^s is the result of replacing (free) occurrences of s (if s is a variable) by t provided that the variables in s (if s is a term other than a variable) which are bound in A are those variables of t which are bound in A_t^s. Otherwise A_t^s is A.

A model for HPC$^=$ will be an ordered pair $\langle D, I \rangle$ such that:

> D is non-empty and I is a function such that
> $I(c_n) \in D$
> $I(F^n) \subseteq D^n$
> $I(f^n)$ is a function from D^n to D.

As before we will use sequences α in defining satisfaction and we require that $\alpha(v) \in D$, $\alpha(c_n) = I(c_n)$, and $\alpha(f^n(t_1, \ldots t_n)) = I(f^n)(\alpha(t_1) \ldots \alpha(t_n))$.

The definition of satisfaction is now exactly the same as for HPC with the additional stipulation that α satisfies $t_1 = t_2$ iff $\alpha(t_1)$ and $\alpha(t_2)$ are identical. This last condition in effect requires that $I(=) = \{\langle x, x\rangle : x \in D\}$. The assignment of any equivalence relation of indistinguishable elements of M would satisfy the identity axioms, but we are only concerned with what are called *normal* models of $\text{HPC}^=$, i.e. those where identity is identity.

EXERCISE 1. Give an example of a formula which is valid only in domains with 3 objects.

EXERCISE 2. Show that the Löwenheim-Skolem theorem as stated for HPC does not hold for $\text{HPC}^=$.

We can prove the completeness of $\text{HPC}^=$ with a slight variation of the constructions and proofs of the earlier section on HPC.

LEMMA. *If Γ is a set of formulas such that:*

(1) $A \in \Gamma$ *iff* $-A \notin \Gamma$
(2) $(A \supset B) \in \Gamma$ *iff* $A \notin \Gamma$ *or* $B \in \Gamma$
(3) $(v)A \in \Gamma$ *iff for all* t, $A_t^v \in \Gamma$
(4) Γ *is* $\text{HPC}^=$ *consistent*

then there is an α and an M, such that the domain of M is a subset of the natural numbers and $A \in \Gamma$ iff α satisfies A in M.

Proof. We can define α and M as before except for the complication that we must ensure that identity comes out right. First, we divide the terms into equivalence classes E_t where $E_t = \{s : (s = t) \in \Gamma\}$. We know that E_t is an equivalence class because the reflexivity, symmetry and transitivity of $=$ are provable in $\text{HPC}^=$ and by clauses (1) and (4) we can show that these formulas are in every such Γ. We assume given some enumeration of the terms of the language $t_1, \ldots t_n, \ldots$ and we define a function μ on the terms such that $\mu(t)$ = the least i such that $t_i \in E_t$. [For example, if x_1 is the first term in our list and $t = x_1 \in \Gamma$, then $\mu(t) = 1$.]

We can now define α, D and I. $D = \{n$: for some t, $\mu(t) = n\}$. (Note that $D \subseteq N$)

$$\alpha(v) = \mu(v)$$
$$I(c_n) = \mu(c_n)$$
$$I(f^n) = \{\langle \mu(t_1), \ldots \mu(t_n), \mu(ft_1 \ldots t_n) > : f^n t \ldots t_n \text{ occurs in } \Gamma\}$$
$$I(F^n) = \{\langle \mu(t_1), \ldots \mu(t_n)\rangle : F^n_t, \ldots t_n \in \Gamma\}.$$

We will now show by induction on the number of function symbols in t that $\alpha(t) = \mu(t)$.

If $k = 0$; if t is a variable v, $\alpha(v)$ is defined as $\mu(v)$. If t is a constant, $\alpha(c) = I(c) = \mu(c)$.

$k = m + 1$. We assume that for t with m function symbols $\alpha(t) = \mu(t)$, and consider a term $ft_1 \ldots t_n$ with $m + 1$ function symbols. By our definition $\alpha(ft_1 \ldots t_n) = I(f)(\alpha(t_1) \ldots \alpha(t_n))$ so by our induction hypothesis $\alpha(ft_1 \ldots t_n) = I(f)(\mu(t_1) \ldots \mu(t_n))$, and the definition of $I(f^n)$ is $\{\langle \mu(t_1), \ldots \mu(t_n), \mu(ft_1 \ldots t_n)\rangle : ft_1 \ldots t_n \in \Gamma\}$ and so $\alpha(ft_1 \ldots t_n) = \mu(ft_1 \ldots t_n)$.

Using this fact we can show that for all atomic $A \in \Gamma$, α satisfies A in M iff $A \in \Gamma$.

Proof. If A is $Ft_1 \ldots t_n$ then by definition α sat $Ft_1 \ldots t_n$ iff $\langle \alpha(t_1), \ldots \alpha(t_n)\rangle \in I(F)$, but we know that $I(F^n) = \{\langle \mu(t_1), \ldots \mu(t_n)\rangle : F^n t_1 \ldots t_n \in \Gamma\}$ and since $\alpha(t_i) = \mu(t_i)$, α sat $Ft_1 \ldots t_n$ iff $Ft_1 \ldots t_n \in \Gamma$.

If A is $t_1 = t_2$, then α sat A iff $\alpha(t_1) = \alpha(t_2)$, but $\alpha(t_1) = \mu(t_1)$, $\alpha(t_2) = \mu(t_2)$ and by definition of μ $\mu(t_1) = \mu(t_2)$ iff $t_1 = t_2 \in \Gamma$.

The proof for nonatomic formulas is exactly the same as the argument for Henkin sets in HPC.

In order to prove completeness it will suffice now to show that any consistent set of formulas can be extended to a Henkin set. I leave it to you to verify that the proof on pp. 19–20 works for HPC$^=$ without any changes at all.

COMPLETENESS THEOREM FOR HPC$^=$. *If A is valid in* HPC$^=$ *then it is provable in* HPC$^=$.

EXERCISE 3. Prove this theorem.

As before we can obtain proofs of compactness and the Löwenheim-Skolem theorem with one modification. The modification results from the fact that we only know that the model is countable rather than knowing that it is countably infinite.

You should notice that the L-S theorem assures us that if a set of formulas has a model then the set has a model with finite or denumerable domain, but the new model does not have any known relation to the old model except that of satisfying the same formulas. We will now prove a stronger version of the theorem which will show that we can find new, 'smaller' models by restricting 'big' models. Before we state the theorem we need a few definitions:

DEFINITION. If S_1 and S_2 are sets, we say that *the cardinality of* S_1 *is at least as great as that of* S_2 ($\bar{\bar{S}}_1 \geqslant \bar{\bar{S}}_2$) iff there is a one-one function from S_2 onto a subset of S_1.

DEFINITION. S_1 *and* S_2 *have the same cardinality* ($\bar{\bar{S}}_1 = \bar{\bar{S}}_2$) iff $\bar{\bar{S}}_1 \geqslant \bar{\bar{S}}_2$ and $\bar{\bar{S}}_2 \geqslant \bar{\bar{S}}_1$.

DEFINITION. *A model* $\langle D_2, I_2\rangle$ *is a restriction of a model* $\langle D_1, I_1\rangle$ *to* D_2 iff for each predicate letter F^n, $I_2(F^n) = I_1(F^n) \cap D_2^n$, and for each function symbol f^n, $I_2(f^n) = I_1(f^n) \cap D_2^{n+1}$ and for all constants c_n, $I_2(c_n) = I_1(c_n) \cap D_2$.

STRONG LÖWENHEIM-SKOLEM THEOREM. *If* Δ *is a set of sentences which are true in* $\langle D_1, I_1\rangle$ *and* $S \subseteq D_1$, *then there is a model* $\langle D_2, I_2\rangle$ *such that*

(1) $S \subseteq D_2 \subseteq D_1$

(2) $\bar{\bar{D}}_2 \leqslant \overline{\overline{S \cup N}}$, $N =$ *the natural numbers*

(3) $\langle D_2, I_2\rangle$ *is the restriction of* $\langle D_1, I_1\rangle$ *to* D_2

(4) *All sentences of* Δ *are true in* $\langle D_2, I_2\rangle$.

Proof. We note first that for any formula A there is a formula A^* in prenex normal form such that A^* is true in M iff A is true in M. [A^* is in prenex normal form iff it is of the form $(Qv_1)\dots(Qv_n)B$

where B contains no quantifiers.] Thus we need only consider sets of prenex formulas.

We will now show that for any formula A^* in prenex form there is a formula A_f with no quantifiers such that A^* is true in $\langle D, I\rangle$ iff A_f is true in $\langle D, I_f\rangle$, where I_f is the result of extending I to include interpretations of some function symbols which were not in the domain of I. That is we add some new function symbols and give a suitable extension of I to interpret them.

We prove the existence of A_f by induction on the number of quantifiers in A^*.

$$n = 0. \quad A_f = A^*.$$

$n = k + 1$. If $A^* = (v)B$ then we let A_f be B_f. That is $(v)B$ is true in a model iff B is, and since B contains at most k quantifiers by *IH* there is a B_f.

If $A^* = (\exists\, v)B$ and $v_1, v_2, \ldots v_n$ are all the free variables of $(\exists\, v)B$, then we show that $B^v_{f(v_1,\ldots v_n)}$ is satisfiable iff $\exists\, vB$ is, where f is a function symbol which does not occur in B. If for some α, α sat $B^v_{f(v_1\ldots v_n)}$ in M then α also sat $\exists vB$ for if we take $\beta \underset{v}{\approx} \alpha$ such that $\beta(v) = \alpha(f(v_1 \ldots v_n))\beta$ sat B. Conversely, if all α satisfy $\exists\, vB$ in M then we know that for each n-tuple $\langle d_1, \ldots d_n\rangle$ in the domain there is at least one d such that if a sequence β assigns $d_1, \ldots d_n$ to $v_1, \ldots v_n$ respectively and $\beta(v) = d$ then β sat B. To define the interpretation of f we choose for each $\langle d_1, \ldots d_n\rangle$ a unique d and assign $\langle d_1, \ldots d_n, d\rangle$ to f. [The choices in question require the Axiom of Choice – in fact our theorem is *equivalent* to the Axiom of Choice.]

Hence, for any A^*, by induction we have shown that a formula A_f exists such that A^* is true in M iff A_f is true in an extension of M. (M is extended to include interpretations of the new function symbols.) To prove our theorem we note that we can extend our language to add an infinite list of new function letters $g_1, \ldots$ and that Δ is true in M iff Δ_g is true in M_g, where Δ_g is the result of replacing each A in Δ by A_g such that no g occurs in more than one A_g and M_g is an extension of M which assigns the required interpretations to the g_i.

If $M = \langle D_1, I_1\rangle$ and $M_g = \langle D_1, I_{1g}\rangle$ and $S \subseteq D_1$, we construct our new model as follows:

let $S_0 = S \cup \{d: d \in D_1$ and for some $c_n I_1(c_n) = d\}$
$S_{n+1} = S_n \cup \{d: d \in D_1$ and for some $\langle d_1, \ldots d_m \rangle \in S_n^m$ and some f or $g I_{1g}(f)(d_1, \ldots d_n) = d$ or $I_{1g}(g)(d_1, \ldots dn\rangle = d.\}$

Note first that S_0 is the result of adding at most countably many elements to S. But by induction each S_n is such that $\overline{\overline{S_n}} < \overline{\overline{S \cup N}}$ for there only denumerably many finite sequences formed from a denumerable set and we add only a denumerable number of elements for each of these sequences. Therefore, if $S_\omega = \bigcup_{n \in \omega} S_n$, $\overline{\overline{S_\omega}} \leqslant \overline{\overline{S \cup N}}$.

We now let $D_2 = S_\omega$ and let I_{2g} be the restriction of I_{1g} to D_2. We show now that all the formulas $A_g \in \Delta_g$ are true in $\langle D_2, I_{2g} \rangle$. Suppose α is a sequence of elements of D_2, and that α sat A_g in $\langle D_1, I_{1g} \rangle$. Since A_g contains no quantifiers whether α satisfies A_g depends only on what α assigns to the terms in A_g and what n-tuples are assigned to the predicate letters; thus since I_{2g} is simply the restriction of I_{1g}, α sat A_g in $\langle D_2, I_{2g} \rangle$. By our lemma above if $A^* \in \Delta^*$ then A^* is true in $\langle D_2, I_{2g} \rangle$ iff A_g is. And finally, since A^* does not contain g_i, for any i, A^* is true in $\langle D_2, I_2 \rangle$ iff A^* is true in $\langle D_1, I_1 \rangle$, where I_2 is the restriction of I_{2g} to the original vocabulary.

CHAPTER V

FIRST ORDER THEORIES WITH EQUALITY

We define *a first order theory* T to be a set of closed formulas of $HPC^=$ and the vocabulary of T is the set of logical symbols and individual variables of $HPC^=$ plus those predicate, function and individual constants that appear in T. *A model of* T is a $HPC^=$ model M restricted to the vocabulary of T and such that all formulas of T are true in M.

A theory T is consistent iff the set of sentences T is consistent in $HPC^=$. By strong completeness, if T is consistent then it has a model. A formula A *is a theorem of* T iff $T \vdash A$.

One interesting property of theories is categoricity. Two models M_1 *and* M_2 *are isomorphic* iff there is a 1-1 function τ such that τ maps D_1 onto D_2 and

(1) $I_2(c_n) = \tau(I_1(c_n))$

(2) $I_2(F^n) = \{\langle \tau(d_1), \ldots \tau(d_n)\rangle : \langle d_1, \ldots d_n\rangle \in I_1(F^n)\}$

(3) $I_2(f^n) = \{\langle \tau(d_1), \ldots \tau(d_n), \tau(d_{n+1})\rangle : \langle d_1, \ldots d_{n+1}\rangle \in I_1(f^n)\}$.

EXERCISE 1. Show that if α satisfies A in M_1 and M_1 and M_2 are isomorphic then there is a β such that β satisfies A in M_2.

A theory is *categorical* iff all of its models are isomorphic.

Although categoricity is an interesting property, very few theories are categorical. Somewhat more useful is the concept of categoricity with respect to the cardinality $\bar{\bar{S}}$. *A theory* T *is categorical with respect to* $\bar{\bar{S}}$ iff for any $\langle D_1, I_1\rangle\langle D_2, I_2\rangle$, if $\bar{\bar{D}}_1 = \bar{\bar{D}}_2 = \bar{\bar{S}}$ then if $\langle D_1, I_1\rangle$ and $\langle D_2, I_2\rangle$ are models of T then they are isomorphic. The idea behind picking out the concept of categoricity or categoricity with respect to $\bar{\bar{S}}$ is that categorical theories completely determine the structure of their models.

A syntactic idea which one might intuitively guess would parallel categoricity would be that of a theory which determines the truth of every closed formula in the language. That is if A is a closed formula, either $T \vdash A$ or $T \vdash A$. Unfortunately, the name that was chosen for this property is *completeness*. A theory T is *theory complete* (T-complete) iff for all closed A, either $T \vdash A$ or $T \vdash -A$.

EXERCISE 2. Show that if a theory T is categorical then it is T-complete.

EXERCISE 3. Show that if a theory T is $\aleph_0$ categorical and has no finite models then it is T-complete.

Notice that we put no restriction on the set of closed sentences T, but in practice we are usually interested in theories which have specific properties of axiomatizability. T is *effectively axiomatizable* iff there is a set S such that $S \vdash A$ iff $T \vdash A$ and there is a decision procedure which decides for any given formula A whether $A \in S$. T is *finitely axiomatizable* iff there is a finite set S such that $T \vdash A$ iff $S \vdash A$.

Returning now to categoricity and completeness, let us consider some examples of mathematical theories. If we are formalizing the theory of total dense orderings with neither first nor last elements, we would hope to find a set of axioms that is both T-complete and categorical.

Let T_d be the theory with non-logical vocabulary F_0^2 and the following axioms: (we will abbreviate Fv_1v_2 as $v_1 < v_2$ to make the axioms more readable).

$(x_1)(x_2)[x_1 < x_2 \vee x_2 < x_1 \vee x_1 = x_2]$	totalness
$(x_1)(x_2)(x_3)[x_1 < x_2 \supset (x_2 < x_3 \supset x_1 < x_3)]$	transitivity
$(x_1)[-(x_1 < x_1)]$	non-reflexivity
$(x_1)(\exists x_2)(\exists x_3)[x_2 < x_1 < x_3]$	openness
$(x_1)(x_3)(\exists x_2)[x_1 < x_3 \supset x_1 < x_2 < x_3]$.	denseness

This theory is almost as good as we hoped – it can be shown that it is $\aleph_0$ categorical and has no finite models so by the exercise above it is T-complete.

If, on the other hand, we are formalizing the concept of a group, we would not expect categoricity or T-completeness because there are many different groups with different structures.

A basic mathematical structure which we *would* hope to be able to formalize would be the natural numbers; that is, we would expect and want to find a theory of the natural numbers which is categorical, complete and effectively axiomatizable. We shall show that any theory which is adequate for number theory does not have these properties. The proof that there is no categorical theory is easy; the proof that there is no T-complete theory will require a lot of work.

THEOREM. (*Existence of non-standard models*). *Let T be a theory whose vocabulary includes f_0^1 and c_0 and which is such that there is a model $\langle N, I\rangle$ of T where N is the set of natural numbers and $I(c_0) = 0$ and $I(f_0^1) = \{\langle m, n\rangle : m + 1 = n\}$. Then there is a denumerable model of T which is not isomorphic to $\langle N, I\rangle$.*

Proof. We construct a new theory T' whose vocabulary consists of the vocabulary of T plus a new constant c_ω. The theory consists of $T \cup C$ where C is the following set of formulas:

$$c_\omega \neq c_0$$
$$c_\omega \neq f_0^1(c_0)$$
$$c_\omega \neq f_0^1 f_0^1 c_0$$
$$\vdots$$

We will abbreviate $\underbrace{f_0^1 \ldots f_0^1}_{n}(c_0)$ by 0^n. Let C' be the set of formulas $0^n \neq 0^m$ where $m \neq n$; all of these are true in $\langle N, I\rangle$. Now we consider the set $T \cup C \cup C'$ and let Δ be any finite subset of that set. Every sentence in $\Delta \cap T$ is satisfied in $\langle N, I\rangle$ as is every sentence in $C' \cap \Delta$. Every sentence in $C \cap \Delta$ is of the form $c_\omega \neq 0^n$. Let k be the largest integer for which $c_\omega \neq 0^k$ is in Δ. We define I' to be I on the vocabulary of T and let $I'(c_\omega) = k + 1$. Every sentence of Δ is true in $\langle N, I'\rangle$. Therefore by the compactness theorem $C \cup T' \cup C'$ has a model and since C' has no finite models the model must be infinite and by the Löwenheim-Skolem theorem we can guarantee that it is countable. Call this model $\langle N, I_2'\rangle$ and let $\langle N, I_2\rangle$ be the restriction to the vocabulary of T.

Suppose now that $\langle N, I_2\rangle$ is isomorphic to $\langle N, I\rangle$ – then there is a g such that $I_2(c_0) = g(0)$ and $I_2(f_0^1) = \{\langle g(n), g(m)\rangle: n+1 = m\}$. Let $I_2(c_\omega) = j$. We know that for some n $j = g(n)$, and so we also know that $I_2(0^n) = j$ since $I(0^n) = n$. But then it follows that $c_\omega = 0^n$ is true in the model, hence not all sentences of C are true in the model. Thus the assumption of $\aleph_0$ categoricity leads to a contradiction.

EXERCISE 4. Show that set theory is not categorical. (Hint: Add constants $c_0, \ldots$ and the sentences $c_{n+1} \in c_n$, and apply compactness.)

EXERCISE 5. Show that if T has models with arbitrarily large finite domains then it has an infinite model.

There is a sense in which function symbols can be dispensed with. We can replace a theory containing an n-place function symbol by a theory containing a new $n+1$ place predicate without loss of expressive power, though we must modify the theory somewhat. Let T_f be a theory containing an n-place function symbol f. We obtain T_F as follows: Each occurrence of f in an atomic formula $Af(t_1 \ldots t_n)$ is replaced by $(\exists v)(A_v^{f(t_1 \ldots t_n)} \wedge Ft_1 \ldots t_n v)$. We add to the theory the sentence $(x_1)(x_2)\ldots(x_n)(x_{n+1})(x_{n+2})$ $[Fx_1 \ldots x_n x_{n+1} \supset (Fx_1 \ldots x_n x_{n+2} \supset x_{n+1} = x_{n+2})]$, to form T_F.

THEOREM ON THE ELIMINATION OF FUNCTION SYMBOLS. *T_f is satisfiable iff T_F is.*

Proof. Suppose $\langle D, I\rangle$ satisfies T_f. We let $\langle D, I^*\rangle$ be such that $I^* = I$ on the common vocabulary and $I^*(F) = I(f)$.

An atomic formula of T_f will be satisfied by α in $\langle D, I\rangle$ iff α satisfies the translation of the atomic formula in T_F. It is easy to show by induction that this property holds for all formulas. The additional sentence will also be satisfied by the construction of I^*.

In the other direction, suppose $\langle D, I^*\rangle$ is a model for T_F. We let $\langle D, I\rangle$ be such that $I = I^*$ on their common vocabulary and

$$I(f) = I^*(F).$$

We know that $I^*(F)$ is a function since the sentence

$(x_1)\ldots(x_{n+2})(Fx_1\ldots x_{n+1} \supset (Fx_1\ldots x_n x_{n+2} \supset x_{n+1} = x_{n+2}))$ is true in $\langle D, I^*\rangle$.

COROLLARY 1. *$T_f \vdash A_f$ iff $T_F \vdash A_F$.*

Proof. If $T_f \vdash A_f$ then A_f follows from some finite set of axioms of T_f; call the conjunction of this set T'_f. Then $T'_f \vdash A_f$ and by soundness $T'_f \models A_f$ so $T'_f \cup \{-A_f\}$ is not satisfiable. Thus $T'_F \cup \{-A_F\}$ is not satisfiable so $T'_F \models A_F$ and by completeness $T'_F \vdash A_F$. Since $T_F \vdash T'_F$, $T_F \vdash A_F$. The proof in the other direction is exactly similar.

We will show next that for theories in $\mathrm{PC}^=$ there is a corresponding theory without identity. Given a theory $T^=$ we form the new theory by replacing each occurrence of $=$ in a $T^=$ axiom by a new two-place predicate letter F_i^2 and for each n-place predicate letter F^n in the theory we add the axioms

$$(x_1)(x_2)\ldots(x_n)(x_{n+1})(F_i^2 x_1 x_{n+1} \supset (F^n x_1 \ldots x_n \supset Fx_{n+1}x_2\ldots x_n))$$

$$\vdots$$

$$(x_1)(x_2)\ldots(x_n)(x_{n+1})(F_i^2 x_n x_{n+1} \supset (F^n x_1 \ldots x_n \supset Fx_1\ldots x_{n-1}x_{n+1}))$$

and we also add the axioms $(x_1)F_i^2x_1x_1$, $(x_1)(x_2)(F_i^2x_1x_2 \supset F_i^2x_2x_1)$, $(x_1)(x_2)(x_3)[F_i^2x_1x_2 \supset (F_i^2x_2x_3 \supset F_i^2x_1x_3)]$.

Let us call the new theory T^F.

Eliminability of identity. $T^=$ is satisfiable iff T^F is satisfiable.

Proof. If $M^=$ is a model for $T^=$ then we can obtain a model M for T^F by letting M be the same as $M^=$ except that it assigns the identity relation to F_i^2.

If M is a model for T^F then since $I(F)$ is an equivalence relation on D we can form equivalence classes E_d for each element of D. $E_d = \{d' : \langle d, d'\rangle \in I(F_i^2)\}$. We construct a new domain D_E consisting of the equivalence classes of D and we let

$$I_E(F^n) = \{\langle E_{d_1}, \ldots E_{d_n}\rangle : \langle d_1, \ldots d_n\rangle \in I(F^n)\}.$$

Note that $I_E(F_i^2) = \{\langle E_{d_1}, E_{d_2}\rangle : \langle d_1, d_2\rangle \in I(F_i^2)\}$, and since $E_{d_1} = E_{d_2}$ iff

$\langle d_1, d_2\rangle \in I(F_i^2)$, $I_E(F_i^2)$ is the identity relation. Finally we let α_E be the sequence that assigns E_d wherever α assigns d.

LEMMA 1. *α_E satisfies A in $\langle D_E, I_E\rangle$ iff α satisfies A in $\langle D, I\rangle$.*
Proof. By induction on the order of A.

EXERCISE 6. Prove Lemma 1.

Since $I_E(F_i^2)$ is the identity relation and $\langle D_E, I_E\rangle$ satisfies the sentences of T^F, we can easily construct a model $\langle D_E, I'_E\rangle$ for $T^=$ by letting I'_E be like I_E except that it is not defined on F_i^2.

COROLLARY. *$T^= \vdash A^=$ iff $T^F \vdash A^F$.*

EXERCISE 7. Prove this corollary.

CHAPTER VI

GÖDEL'S INCOMPLETENESS THEOREMS: PRELIMINARY DISCUSSION

In this section I will try to outline the ideas behind the proofs and indicate what facts need to be established about first order theories which are intended to formalize number theory in order to prove the theorems. The first argument will be a non-constructive version of the T-incompleteness of number theories – non-constructive in the sense that it does not produce a particular formula A which is such that neither A nor $-A$ is a theorem, but only shows that some such A exists.

We will use the fact that there is a three place predicate $T(x, y, z)$ and an effective mapping from computable functions and computations into the natural numbers such that $T(m, n, k)$ is true iff k is the number of a computation of the value of the mth function for argument n. (In order to obtain the mapping from all computable functions it is necessary to include partial functions, i.e. ones which are undefined for some arguments.) We will say that a decision problem of the form 'Is $n \in S$' is solvable or has a positive solution iff there is a total computable function $f(x)$ such that $f(n) = 0$ if $n \in S$ and $f(n) = 1$ if $n \notin S$. Consider the question whether for a particular value of n the nth computable function is defined for the argument n, i.e., whether $(Ez)T(n, n, z)$ – is there a decision procedure for this problem? If there is a positive solution then that means there is a total computable function $g(x)$ such that $g(n) = 0$ if $(Ez)T(n, n, z)$ and $g(n) = 1$ if $(z) - T(n, n, z)$. Let us define the function $h(x)$ so that $h(x) = 0$ if $g(x) = 1$ and $h(x)$ is undefined otherwise. Clearly h is computable if g is, so if g exists there must be a number n such that h is the nth function in our enumeration. By the characteristic property of g, $h(n)$ is defined iff $g(n) = 1$, which holds iff $(z) - T(n, n, z)$ which is true iff $h(n)$ is undefined. Therefore the assumption that g is total, computable and a decision procedure for whether $(Ez)T(n, n, z)$ leads to a contradiction; therefore there is no solution.

All of the above argument is typical informal mathematical proof; to obtain from it a metatheorem about first order theories we need to establish some connections with provability in first order theories. We shall show that for any adequate[1], consistent first order number theory there is a formula $\underline{T}(\underline{x}_1, \underline{x}_2, \underline{x}_3)$ such that $\underline{T}(\underline{m}, \underline{n}, \underline{k})$ is provable in the theory iff $T(m, n, k)$.[2] [We will often be discussing object language formulas which intuitively represent various metalanguage predicates and functions so for the sake of perspicuousness I will use underlined letters, e.g., $\underline{f}$ to stand for the object language symbol or formula or function which represents f. For example, $\underline{17}$ is the name of the object language symbol consisting of seventeen iterations of successor symbol applied to the constant which is interpreted as zero.] It is also easy to show that whenever $\underline{T}(\underline{m}, \underline{n}, \underline{k})$ is provable then so is $(\underline{Ex}_1)\underline{T}(\underline{m}, \underline{n}, \underline{x}_1)$ and so we know that this last formula will be provable whenever $(Ez)T(m, n, z)$. We will also show that there is an effective enumeration of the theorems of any effectively axiomatizable first order theory. Suppose now that we had a consistent, adequate effectively axiomatizable first order number theory. Then we could enumerate the theorems and we would eventually find for each n either $(\underline{Ex}_1)\underline{T}(\underline{n}, \underline{n}, \underline{x}_1)$ or $(\underline{x}_1) - \underline{T}(\underline{n}, \underline{n}, \underline{x}_1)$, and this would give us a decision procedure for $(Ez)T(n, n, z)$ unless the formal system were somehow incorrect. How could it be incorrect? We already know that $(\underline{Ex}_1)\underline{T}(\underline{n}, \underline{n}, \underline{x}_1)$ is provable when $(Ez)T(n, n, z)$, and if both that formula and its negation were provable the system would be inconsistent. Thus there is left only the possibility that $(\underline{Ex}_1)\underline{T}(\underline{n}, \underline{n}, \underline{x}_1)$ is provable even though it is not the case that $(Ez)T(n, n, z)$. But if so then $T(n, n, k)$ is always false and so $-\underline{T}(\underline{n}, \underline{n}, \underline{k})$ is provable for each k. Thus the system will have as theorems formulas $(\underline{Ex}_1)Ax_1$ and $-\underline{A0}$, $-\underline{A1}, \ldots$ Such a system is said to be ω-inconsistent. Thus we can conclude that any adequate, ω-consistent effectively axiomatizable first order number theory is not complete.

In order to give a constructive proof of this theorem in which we find a specific closed formula A such that neither A nor $-A$ is a theorem, we must use a different method. Consider first the possibility that some formula A of the system might be such that it somehow expressed the statement: A is false. Such a formula is true if it is false and false if it is true. Thus any system in which such a statement is expressible must be inconsistent. (We will later give a more formal version of this argument which will show that the truth predicate for a theory is not definable in that theory.) The self-reference involved in a sentence need not, of course, be as blatant as the case above, for example:

> The sentence on lines 9 and 10 of page 51 of *Advanced Logic for Applications* is false.

Any such sentence will lead to paradox.

Gödel's remarkable idea is that if we substitute 'unprovable' for 'false' in such sentences we do not get a paradox, but instead a surprising metatheorem. Consider then the possibility that we could find a sentence of a formal theory T which expressed the following sentence S:

> S is not provable in T.

If S is provable, then there is a false theorem in T; if S is not provable then there is a true but unprovable sentence of T. In either case T is defective. Of course it is far from obvious that any formula such as S exists in any reasonable system, and so the largest part of the work to be done in proving the theorem consists in showing that there is such a formula. We will now give a slightly more detailed outline of the proof, still making large assumptions about the existence of object language formulas which correspond to metalanguage sentences.

We assume that there are effective mappings of formulas and sequences of formulas into the natural numbers such that there is a formula $\underline{\mathrm{Pr}}(x_1, x_2)$ and a functional expression $\underline{\mathrm{Sub}}(x_1, x_2)$ such that the following conditions are satisfied:

(Ia) $\vdash \underline{\mathrm{Pr}}(\underline{m}, \underline{n})$ iff m is the number of a proof of formula number n

(Ib) $\vdash -\underline{\mathrm{Pr}}(\underline{m}, \underline{n})$ iff m is not the number of a proof of formula number n

(IIa) $\vdash \underline{\mathrm{Sub}}(\underline{m}, \underline{n}) = \underline{k}$ iff k is the number of the formula which results from substituting the numeral $\underline{m}$ for the variable $\underline{x}_1$ in the formula whose number is n

(IIb) $\vdash \underline{\mathrm{Sub}}(\underline{m}, \underline{n}) \neq \underline{k}$ iff k is not the number of a formula which results from substituting the numeral $\underline{m}$ for the variable $\underline{x}_1$ in formula n.

We will define a theory to be *ω-consistent* if whenever a formula of the form $(\underline{E}\underline{x}_1)\underline{A}\underline{x}_1$ is provable it is not the case that all the formulas $-\underline{A0}$, $-\underline{A1}, \ldots$ are provable. If a system is ω-consistent, then it is consistent. If it is not ω-consistent then it is *ω-inconsistent.*

Since $\underline{\mathrm{Pr}}(\underline{x}_2, \underline{x}_1)$ is a formula, then so is its negation and so also is $(\underline{x}_2) - \underline{\mathrm{Pr}}(\underline{x}_2, \underline{x}_1)$ so if we substitute the functional expression $\underline{\mathrm{Sub}}(\underline{x}_1, \underline{x}_1)$ for $\underline{x}_1$, we obtain the formula $(\underline{x}_2) - \underline{\mathrm{Pr}}(\underline{x}_2, \underline{\mathrm{Sub}}(\underline{x}_1, \underline{x}_1))$. Let the number assigned to this formula in our effective enumeration be n. Consider the formula which results if we substitute the numeral $\underline{n}$ for $\underline{x}_1$ in the formula n, i.e. the formula $(\underline{x}_2) - \underline{\mathrm{Pr}}(\underline{x}_2, \underline{\mathrm{Sub}}(\underline{n}, \underline{n}))$, and let us call the number of this formula k. Since this is the formula which results from substituting $\underline{n}$ for $\underline{x}_1$ in the formula n, by Assumption IIa, $\vdash \underline{\mathrm{Sub}}(\underline{n}, \underline{n}) = \underline{k}$.

Suppose that the formula $(\underline{x}_2) - \underline{\mathrm{Pr}}(\underline{x}_2, \underline{\mathrm{Sub}}(\underline{n}, \underline{n}))$ is provable. If it is provable then there is a proof and that proof is assigned a number in our enumeration – suppose it is m. By (Ia) we know that $\vdash \underline{\mathrm{Pr}}(\underline{m}, \underline{k})$ and so we know that $\vdash \underline{\mathrm{Pr}}(\underline{m}, \underline{\mathrm{Sub}}(\underline{n}, \underline{n}))$, from which it follows that $\vdash (\underline{E}\underline{x}_2)\underline{\mathrm{Pr}}(\underline{x}_2, \underline{\mathrm{Sub}}(\underline{n}, \underline{n}))$ and so the system is inconsistent. Therefore, *if the system is consistent*, then $(\underline{x}_2) - \underline{\mathrm{Pr}}(\underline{x}_2, \underline{\mathrm{Sub}}(\underline{n}, \underline{n}))$ is not provable.

Thus if the system is consistent no number is the number of a proof of that formula and so by (Ib), $\vdash - \underline{\mathrm{Pr}}(\underline{0}, \underline{\mathrm{Sub}}(\underline{n}, \underline{n}))$, $\vdash - \underline{\mathrm{Pr}}(\underline{1}, \underline{\mathrm{Sub}}(\underline{n}, \underline{n})), \ldots$ for all numerals. Therefore if the system is ω-consistent, then $(\underline{E}\underline{x}_2)\underline{\mathrm{Pr}}(\underline{x}_2, \underline{\mathrm{Sub}}(\underline{n}, \underline{n}))$ is not provable either.

Since we know that ω-consistency entails consistency, we can put together the two facts as:

GÖDEL'S FIRST INCOMPLETENESS THEOREM. *If the system is ω-consistent and adequate for arithmetic and effectively axiomatizable, then neither the formula $(\underline{x}_2) - \underline{\mathrm{Pr}}(\underline{x}_2, \underline{\mathrm{Sub}}(\underline{n}, \underline{n}))$ nor its negation is provable.*

We will now set out two specific number theoretic systems and prove these theorems for them. We will change the basis of the system slightly in order to simplify certain notation matters. Our individual variables will be the infinite list a, $a\#$, $a\#\#, \ldots$ and the language will have one individual constant c_0, one one-place functional constant f_0^1, and three two place constants f_0^2, f_1^2, and f_2^2. For the sake of familiarity and readability we will abbreviate $a\#$ by b, $a\#\#$ by c, and so on. We will also abbreviate c_0 as 0, $f_0^1(a)$ as a', $f_0^2(a, b)$ as $a + b$, $f_1^2(a, b)$ as ab, and finally $f_2^2(a, b)$ as a^b. Our system will have the five axiom schema of quantification theory and the axiom schemata and axiom of identity. We will add axioms corresponding to the Peano postulates and some axioms characterizing the properties of the functional constants. The system N will also have the axiom schema of induction, whereas Q will have one additional axiom in place of the induction schema. Thus Q is the result of adding only a finite number of axioms to the logical axioms – this will be important when we prove the unsolvability of the decision problem for quantification theory.

The two rules of both systems will be modus ponens and the generalization rule of HPC$^=$, and the definitions of term, well-formed formula, free and bound occurrences of variables, and so on, are the definitions of HPC$^=$ restricted to the vocabulary of our theories. [Note that, for example, there are no predicate letters in the language.]

(1) $A \supset (B \supset A)$
(2) $(A \supset B) \supset [(A \supset (B \supset C)) \supset (A \supset C)]$
(3) $(-A \supset -B) \supset (B \supset A)$
(4) $(v)[A \supset B] \supset [A \supset (v)B]$, if v is not free in A
(5) $(v)A \supset A_t^v$
(6) $(a)(a = a)$
(7) $(a)(b)[a = b \supset [A \supset A_b^a]]$
(8) $(a)[a' \neq 0]$
(9) $(a)(b)[a' = b' \supset a = b]$
(10) $(a)(Eb)[a \neq 0 \supset a = b']$
(11) $(a)[a + 0 = a]$
(12) $(a)(b)[a + b' = (a + b)']$

(13) $(a)(b)[a(b') = ab + a]$

(14) $(a)[a0 = 0]$

(15) $(a)[a^0 = 0']$

(16) $(a)(b)[a^{b'} = a^b b]$

(17N) $A^a_0 \supset [(a)(A \supset A^a_{a'}) \supset (a)A$

(17Q) $(a_1)(a_2)(a_3)(a_4)(a_5)(a_6)[(a_2 = a_1 a_3 + a_4 < a_1 \wedge a_2 = a_1 a_6 + a_5 \wedge a_5 < a_1) \supset a_4 = a_5]$, where $a < b$, is an abbreviation for $(Ec)[c + a' = b]$, $(Ev)A$ is an abbreviation for $-(v) - A$ and $A \wedge B$ is an abbreviation for $-(A \supset -B)$.

NOTES

[1] 'Adequate' means roughly that addition, multiplication and exponentiation are represented – we will define it precisely later.

[2] We can also show that $-\underline{T}(\underline{m}, \underline{n}, \underline{k})$ is provable iff $-T(m, n, k)$.

CHAPTER VII

UNDECIDABILITY AND INCOMPLETENESS

Our main objectives are to show

(I) $Q(N)$ is incomplete, i.e. there is a closed A such that neither $\vdash_Q A$ nor $\vdash_Q - A$.

(II) $Q(N)$ is undecidable, i.e. there is no effective way of deciding whether $\vdash_Q A$.

Using (II) for Q we can establish

(III) Functional calculus is undecidable.

(I) and (II) would only show that Q and N are inadequate formalizations of our intuitive concepts if we could not also show

(IV) Any consistent effective extension of Q is incomplete.

(V) Any consistent effective extension of Q is undecidable.

Given the theorem proved earlier that if a theory T is complete then it is decidable we could use (II) to establish (I) and (V) to establish (IV), but it is desirable to show (I) independently of (II) in order to keep (I) independent of any notion of effectiveness. Furthermore, applying (II) and (V) gives only non-constructive proofs of (I) and (IV) whereas our methods will produce specific formulas which are neither provable nor disprovable.

In order to establish (II) and (V) rigorously we will need a definition of effectiveness. We first define the notion of 'numeralwise representability in theory T', the idea behind the definition being that many predicates can be expressed in a theory which are not explicitly provided for in the vocabulary.

We will use letters P, R, etc. for informal predicates in the metalanguage and A, B, etc. for formulas in Q or N. When we define a particular predicate, e.g., Prime(x) we will use the same expression ambiguously for the informal predicate and the object language formula which represents it. We will use the letters m, n, k, etc. for specific numbers in the object language and $\underline{m}, \underline{n}, \underline{k}$, etc. for the corresponding numerals in the object language. We will use the variables x, y, z, etc. as number variables in the metalanguage and as names of arbitrary variables of the object language. The official variables of the object language will be $a, a\#, a\#\#, \ldots$ but we will rarely have occasion to mention specific variables; thus, for example, we will speak of a formula $A(x, y)$ which refers ambiguously to $A(a, \#a)$ or $A(a, a\#\#)$ or $A(a\#, a)$, etc. We could eliminate all of the above ambiguities by adopting enough different notation.

DEFINITION. A predicate $P(x_1, x_2, \ldots, x_n)$ is *numeralwise representable in theory T* iff there is a formula $A(a_1, a_2, \ldots, a_n)$ of T such that $A(\underline{m}_1, \ldots, \underline{m}_n)$ is provable in T iff $P(m_1, \ldots, m_n)$ is true and for any n-tuple of numerals either $A(\underline{m}_1, \ldots, \underline{m}_n)$ or its negation is provable in T.

For our purposes we will assume that *a predicate is effective* iff it is numeralwise representable in Q. Justification of the assumption would consist of showing that numeralwise representability in Q is coextensive with Turing computability and/or general recursiveness. The proof sketched in Ch. V that if a system is complete it is decidable can be used to show that if a system is complete with respect to a formula $A(x)$, i.e., for every n either $A(\underline{n})$ or $-A(\underline{n})$ is provable, then $A(x)$ is decidable. Hence, if a predicate is numeralwise representable in Q we can decide the predicate by deciding the provability of $A(x)$. Therefore any predicate n.r. in Q is effective. The converse will become more plausible as we show various predicates are numeralwise representable in Q.

There is a weaker notion, which we might call weak representability in T, which is defined as follows: A predicate $P(x_1, x_2, \ldots)$ is *weakly n.r.* in T iff there is a formula $A(x_1, x_2, \ldots x_n)$ such that $A(\underline{m}_1, \underline{m}_2 \ldots \underline{m}_n)$ is provable in T iff $P(m_1, \ldots m_n)$ is true. We will show later that a predictate is n.r. iff both it and its negation are weakly representable.

THEOREM 1. *If P is n.r. in T then P is n.r. in any consistent extension of T.*

COROLLARY. *If P is n.r. in Q then it is n.r. in N.*

We wish ultimately to show that the predicate '$A_1, A_2, \ldots, A_n$ is a proof in Q' is effective. In order to do this we must provide a translation of statements about the syntax of Q into statements about numbers. Before doing this, however, we develop some general theorems about numeralwise representability in Q.

(1) $m' = n, m = n + k, m = n \cdot k$, and $m^n = k$ are n.r.

Proof. Verify that the appropriate axioms provide for derivations of the corresponding equation if it is true; if, e.g., $m = n \cdot k$ is false, then $\underline{m} \neq \underline{j}$ and $\underline{j} = \underline{n} \cdot \underline{k}$ must be provable for some j, hence $\underline{m} \neq \underline{n} \cdot \underline{k}$ is provable.

(2) If $P(x)$ and $R(y)$ are n.r., then so are $P(x) \vee R(y)$, $P(x) \wedge R(y)$, $P(x) \supset R(y)$ and $-P(x)$.

Proof. By obvious facts about provability of truth functional compounds of sentences.

(3) $m < n$ is n.r. by $(Ew)\ (w + m' = n)$.

If true the formula is provable by (1) and E introduction. If false, then $m = n + k$ for some k. $-(w + k' = 0)$ is provable with free variable w since $w + k' = (w + k)' \neq n$ will be provable with free w which can be generalized.

(4) For each n, $(x)[x = 0 \vee \cdots \vee x = n \vee -(x < n')$ is provable.

Proof. By induction in the metalanguage using Axioms 8, 11 and 12.

(5) If $P(x, y)$ is n.r. by $A(x, y)$, then
(a) $(Ex < w)P(x, y)$ is n.r. by $(Ex)[x < w \wedge A(x, y)]$
(b) $(x < w)P(x, y)$ is n.r. by $(x)[x < w\ A(x, y)]$
(c) $z =$ the least y less than w such that $P(y, x)$ is n.r. by $[A(z, x) \wedge z < w \wedge (v < z) - A(v, x)]$
(d) If $(x)(Ey)Pxy$ and Pxy are n.r., then $z = \mu_y P(x, y)$, read 'z is the least number such that $P(x, z)$', is n.r.

Proof. (a) and (b) can be proved from the n.r.-ability of P and (4) above. (c) follows from (b) and (2) above. (d) then follows from (b) and (2) similarly; note that it is essential that for every x there is a y such that Pxy which guarantees that we can find an appropriate z and use it to disprove any false numerical statements of the relevant form.

(6) 'n divides m evenly' (symbolized n/m) is n.r. by

$(Ex < m')[x/n = m]$

(7) Prime(x) is n.r. by $(z < x)[z/x \supset z = 1] \wedge x > 1$

(8) 'z is the ith prime' (symbolized $z = P_i$) is n.r. by Prime(z) $\wedge$ $(Ex < z^{i^i})[x = \mu y((w < z')(\text{Prime}(w) \supset w/y) \wedge (j < i)(v < w)[\text{Prime}(v) \quad \wedge \quad v^j/y \supset w^{j+1}/y)] \wedge [z^i/x \quad \wedge \quad - z^{i+1}/x]$.

Proof. The complicated expression n.r. something since it is composed out of n.r. predicates in ways that preserve n representability (by 1–7). That it picks out the predicate we want can be seen by examining the strategy involved. We want to find a sequence $2^1, 3^2, \ldots, p_i^i$; given such a sequence $z = p_i$ can be determined. The strategy is to pick out the appropriate sequence using only those predicates we already have. Such a sequence must include every prime up to the last one and their exponents must be increasing; the smallest such sequence is the one we want, so we use the least number operator. We must put some upper bound on the size of such a sequence number in order to use the bounded least number operator. Since such a sequence has i terms the biggest of which is z^i the number of the sequence must be less than $(Z^i)^i$

(9) $x - y = z$ is n.r. by $[y + z = x] \vee [z = 0 \wedge x < y]$.

It will be useful to have functional terms for functions other than those directly symbolized in the system, i.e., successor, addition, multiplication and exponentiation. We will introduce a defined notation for any n.r. function. A function $f(x)$ is n.r. iff the predicate $f(x) = y$ is n.r. in the system. For example, the function P_i, the ith prime, is a n.r. function by (8).

(10) If $f(x)$ is a total n.r. function, and $P(y, z)$ is a n.r. predicate, then $P(f(x), z)$ is a n.r. predicate.

Proof. Let Axy n.r. Rxy, the characteristic predicate of $f(x)$ and let $B(yz)$ n.r. Pyz; then $B(z, w)$ & $z = \mu_y A(x, y)$ n.r. $P(f(x), w)$.

COROLLARY. *If $f(x)$ and $g(x)$ are n.r. functions, then $f(g(x))$ is n.r.*

EXERCISE 1. Prove this corollary.

(11) 'z is the exponent of the ith prime in x', symbolized as $z = (x)_i$, is n.r. by the formula $(Ew < x)[w = P_i \wedge w^z/x \wedge -w^{z+1}/x]$

(12) 'y is the least prime in x', symbolized as $\mathrm{lp}(x) = y$, is n.r. by $(Ei < x)[y = P_i \wedge (x)_i \neq 0 \wedge (j < i)(x)_j = 0]$

(13) 'y is the greatest prime in x', symbolized as $\mathrm{gp}(x) = y$, is n.r. by $(Ei < x)[(x)_i \neq 0 \wedge y = P_i \wedge (j < x)(j > i \supset (x)_j = 0]$

(14) 'w is the length of sequence x' symbolized as $\mathrm{lh}(x) = w$, is n.r. by the expression $w = \mathrm{gp}(x) - \mathrm{lp}(x) + 1$

(15) z is the result of putting sequence y after sequence x will be symbolized as $z = x^*y$ and this relation is n.r. by the formula

$$(i \leqslant \mathrm{gp}(z))[i \leqslant \mathrm{gp}(x) \supset (x)_i = (z)_i \quad \wedge \quad (j \leqslant \mathrm{lh}(y))[(y)_{\mathrm{lp}(y)+j} = (z)_{\mathrm{gp}(x)+j+1}]]$$

(16) A FORM OF RECURSION THEOREM. *If $P(x)$ and $R(y, z, w)$ are* n.r. *then 'Every exponent of z either satisfies P or there are exponents of smaller primes to which it bears R' is n.r.*

Proof. $(i < \mathrm{lh}(z))[A((z)_i) \vee (Ej,k < i)\, B((z)_j, (z)_k, (z)_i)]$ represents the predicate quoted, if A n.r. P and $B(x, y, z)$ n.r. $R(x, y, z)$.

This theorem will be one of the main tools used in arithmetizing the syntax of Q since term, formula and proof are all defined by recursion. The theorem holds, of course, for relations with more than three arguments.

COROLLARY. *Let $R(x)$ be defined as follows*: (a) *If* $(P(x)$ *then* $R(x)$, (b) *If* $R(x)$, $R(y)$ *and* $S(x, y, z)$ *then* $R(z)$, (c) $R(z)$ *only if* $R(z)$ *follows from clauses* (a) *and* (b). *Then if P and S are n.r. and we can find an n.r. function* $g(x)$ *which gives an upper bound for any x on the number of the sequence necessary to establish* $R(x)$, *then* $R(x)$ *is n.r.*

Proof. Note first that $g(x)$ provides an upper bound on both the length of the necessary sequence and on the numbers appearing in the sequence. Let $A(x)$ n.r. $R(x)$ and $B(x, y, z)$ n.r. $S(x, y, z)$. Then the formula $(Ew < g(x))(i < \text{lh}(w)A((w)_i) \vee (Ej,k < i) \;\; B((w)_j, \;(w)_k, \;(w)_i)$ will n.r. $P(x)$.

EXERCISE. The class of primitive recursive functions is intuitively a subclass of the effectively calculable functions. Show that every two-place primitive recursive function is n.r.

> f is a two-place primitive recursive function if
>
> (1) f is a constant function, the successor function or one of the functions $f(x) = x$, $f(x, y) = x$ or $f(x, y) = y$, or (2) $f(x, y) = g(h_1(x, y), h_2(x, y))$ or, (3) there are functions $g(x)$ and $h(x, y)$ such that $f(x, 0) = g(x)$
> $$f(x, y') = h(x, f(x, y)).$$

We now define a correspondence between the primitive symbols and natural numbers.

(	)	–	⊃	∀	=	0	′	+	·	exp	*a*	#
7	11	13	17	19	23	29	31	37	41	43	47	53

DEFINITION. Var(x) iff $(Ew < x)[x = 47 \cdot 53^w]$ 'x is the number of a variable.'

DEFINITION. Num(x) iff $(Ew < x)[x = 29 \cdot 31^w]$ 'x is the number of a numeral.'

We wish to represent terms as sequences of symbols having the appropriate formation properties; looking at the definition of a term we see that t is a term if t is a numeral or variable or if it is composed from terms by ′, +, · or exp. If we can show that these relations and predicates are n.r. and can find an upper bound on the sequence of the parts of a term that we can invoke the recursion theorem to show that Term(t) is n.r.

THEOREM. *Term(t) is n.r.*

Proof. For the recursion theorem let $A(x)$ be $x = 5^w$ & [Var(w) ∨ Num(w)], let $R(x, y, z)$ be $z = x{*}_2^{31} \vee z = 5^7{*}x{*}2^{37}{*}y{*}2^{11} \vee z = 5^7{*}x{*}2^{41}{*}y{*}2^{11} \vee z = 5^7{*}x{*}2^{43}{*}y{*}2^{11}$.

If t is the GN (Gödel number) of a term then there is a sequence of smaller terms each of which is composed out of earlier terms via R or is a numerical or variable, and the last element of the sequence is t. Therefore the GN of such a sequence consists of an increasing sequence of primes with increasing exponents whose last term has exponent t. Since each step of the sequence increases by at least 1 the number of symbols in the term involved, the length of the sequence is at most the length of t, hence is less than t itself. Therefore the sequence consists of less than t elements each of which is less than p_t^t, so the sequence number is less than $(p_t^t)^t$.

> Atf(x) iff $(Ew, y < x)[x = y{*}2^{23}{*}w \wedge \text{Term}(y) \wedge \text{Term}(w)]$
> Atf(x) n.r. and expresses that x is the GN of an atomic formula.

The definition of a well-formed formula parallels that of term and the definition of wff(x) and the proof that it is n.r. will be parallel to the proof for Term(x). We treat a formula as a sequence of symbols and code the sequence as the exponents of successive primes beginning with 3. We show that wff(x) is n.r. by applying the recursion theorem to the definitions of A = Atf(x) and 'z is an immediate superformula of y and x' for R. R can be explicitly defined as

$$z = 3^{13}{*}x \vee z = 3^7{*}x{*}2^{17}{*}y{*}2^{11} \vee z = 3^7{\cdot}5^{19}{\cdot}7^v{*}x{*}2^{11} \wedge \text{Var}(v).$$

By the same argument as was used for termhood, x is the GN of a term only if there is a formation sequence whose number is less than $(p_x^x)^x$.

In order to characterize the axioms and rules of inference of Q and N we still need a predicate which determines whether a particular variable is free and whether a term is free for some variable in a given wff and we need a characterization of the result of substituting a particular term for a free variable in a formula.

To make the description of the predicates and relations more perspicuous and to avoid repeating 'the GN of formula A' or 'the formula whose GN is x' we let 'A' be the GN of formula A.

> Bound(v, i, 'A') = the variable whose number is v is the ith symbol in A and is bound
> Bound(v, i, 'A') is n.r. by the predicate $v = (\text{'}A\text{'})_i \wedge (Ex, y, z < \text{'}A\text{'})[\text{'}A\text{'} = x*y*z \wedge \text{wff}(y) \wedge (Ew < y)\ y = 3^7 \cdot 5^{19} \cdot 7^v * w \wedge \text{lh}(x) < i < \text{lh}(x*y)]$
> Free(v, i, 'A') = df. $\text{Var}(v) \wedge (\text{'}A\text{'})_i = v \wedge -\text{Bound}(v, i, \text{'}A\text{'})$
> Free(v, i, 'A') is n.r.

The definition of 'B is the result of substituting t for all free occurrences of x in A' will be given by defining the sequence of formulas which results from A by performing the substitution on one occurrence of x at a time. The complexity of this definition of substitution is compelled by the fact that we have to deal with formulas having arbitrarily many free occurrences of a particular variable.

Subst('B', t, v, 'A') = '*B is the result of substituting the term whose* GN *is t for the variable whose* GN *is v at all free occurrences of that variable in A' is n.r.*

Proof. We will show that the relation between A, A_i, and A_{i-1} of 'A_i being the result of substituting t for the last occurrence of v in A_{i-1} which is an unsubstituted occurrence left from A' is an n.r. relation. Given this relation B is the desired formula if it is the last in a sequence of formulas beginning with A and each of whose terms has t substituted for the last occurrence of v not yet substituted for. This sequence can have at most as many terms as A has symbols, i.e. lh('A') and the largest exponent is either A or B, hence is less than 'A' + 'B'. Thus the requisite sequence number is

'A_i is the result of substituting t for v at the last free occurrence in A_{i-1} which is identical with an occurrence in A' is n.r. by the predicate

$$(Ex, y, z, w < \text{'}A\text{'})[\text{'}A_i\text{'} = x*t*y \wedge \text{'}A_{i-1}\text{'} = x*2^v*y \wedge \text{'}A\text{'} = x*2^v*z \wedge \text{Free}(v, \text{lh}(x)+1, \text{'}A\text{'}) \wedge -(Ew_1, w_2, w_3 < \text{'}A\text{'})[\text{'}A_{i-1}\text{'} = x*2^v*w_1*w_2 \wedge \text{'}A\text{'} = x*2^v*w_1*w_3 \wedge (Ei \leqslant \text{gp}(x*2^v*w_1))[i < \text{gp}(x*2^v) \wedge \text{Free}(v, i, \text{'}A_{i-1}\text{'})]] \wedge \text{wff}(A_i).$$

We also need a definition for the predicate 't is free for v in A'. This can most easily be obtained by checking that the result of substituting any variable occurring in t for v is free. Thus,

$$\text{FREE}(t, v, \text{'}A\text{'}) \text{ is n.r. by the predicate } (w < t)[\text{Var}(w) \wedge (Ei < \text{lh}(t)) \quad ((t)_i = w \wedge (z < (p^{\text{'}A\text{'}})^{\text{'}A\text{'}})[(j < \text{lh}(z))[(z)j = (\text{'}A\text{'})j \vee (\text{'}A\text{'})j = v \wedge (z)j = w]] \supset (k < \text{gp}(z)) (\text{Free}(w, k, z) \leftrightarrow \text{Free}(v, k, \text{'}A\text{'}))).$$

It is important to distinguish FREE(t, v, 'A') which holds if t is free for v in 'A' from Free(v, k, 'A') which holds if v is free as the kth symbol in 'A'. It will be useful to have a substitution function Sub(t, v, 'A') whose value is the result of substituting t for v at all free occurrences in A. We can get this immediately from Subst if we can establish that for any numbers m, n, k, there is an h such that Subst(h, m, n, k). We therefore note that if k is not the number of a formula or v is not free in 'A' the result of the substitution under our definition is k itself. Therefore Sub(t, v, 'A') is n.r. by fact (10) above.

We can now define Axiom (y), the predicate which is true of all and only those numbers which are GNs of axioms. For Axioms 6 and 8–16 this is trivial since they are single axioms; Axioms 1–5 and 7 and 17N are schema however and require a bit more work. We will give below the definitions for schema 1 and 5.

AXIOM_1 (x) is n.r. by the predicate $(Ez, w < x)[\text{wff}(x) \wedge x = 3^7 * z * 2^{17} \cdot 3^7 * w * 2^{17} * z * 2^{11} \cdot 3^{11}]$

AXIOM_5 (x) is n.r. by the predicate $(Ev, z, t, w < x)[x = 3^7 \cdot 5^7 \cdot 7^{19} \cdot 11^v * z * 2^{11} \cdot 3^{17} * w * 2^{11} \wedge \text{Term}(t) \wedge \text{Free}(t, v, z)$ $\text{wff}(x)$ $\text{Subst}(w, t, v, z)]$.

Similarly, the relation '*B* is the result of a modus ponens inference from *A* and *C*' and '*A* is the result of a quantifier inference from *B*' are n.r. relations.

Given that the predicate Axiom (x) and the relation ImCon(x, y, z) "read '*x* is an immediate consequence of *y* and *z* by modus ponens or of *y* by the quantifier rule' " are n.r. we can define Bew(x, y), read '*x* is the GN of a proof of the formula whose GN is *y*' as

$$(i < \text{lh}(x))][\text{Axiom}\ ((x)_i) \vee (Ej,k < i)\ \text{ImCon}((x)_i,\ (x)_j,\ (x)_k)].$$

Therefore, Bew(x, y) is n.r.

GÖDEL'S FIRST INCOMPLETENESS THEOREM FOR *Q*. *If Q is omega-consistent, then there is a formula* $(b) - \text{Bew}(b, \underline{m})$ *which is neither provable nor refutable in Q.*

DEFINITION. *Q* is omega-consistent iff there is no formula $A(x)$ such that $(Ey)Ay$, $-A(\underline{0})$, $-A(\underline{1}), \ldots$ are all provable.

Strategy of the proof. We wish to construct a formula whose GN is *m* and which represents the assertion 'formula *m* is unprovable'. If such a formula were provable then we could show in *Q* that it is provable by the n.r. of proof, and hence the system would be inconsistent since we could show of *m* that it is both provable and unprovable. If, however, it is disprovable, then the system is omega-inconsistent since the above argument shows that for any *n* that *n* is not a proof of *m*.

Proof. We wish to consider a specific formula, so let *a*, *b*, be the first two variables of *Q* (i.e., *a*, *a* #). Consider the formula $(b) - \text{Bew}(b, \text{Sub}(29 \cdot 31^a, 47, a))$. This formula expresses that no number is the GN of a proof of the formula whose GN is the result of substituting the ath numeral for the variable whose GN is 47 in the formula whose GN is a. This formula has a GN, let it be k, and now consider the formula which results from substituting the numeral $\underline{k}$ for the variable a in formula k, i.e., $\text{Sub}(29 \cdot 31^k, 47, k)$. Since k is the formula above $\text{Sub}(29 \cdot 31^k, 47, k)$ will be the formula $(b) - \text{Bew}(b, \text{Sub}(29 \cdot 31^{\underline{k}}, 47, \underline{k}))$. Let m be the number of this formula (which we will baptize Kurt for brevity); we know that Sub is a n.r. function and so we can prove in Q that $\underline{m} = \text{Sub}(29 \cdot 31^{\underline{k}}, 47, \underline{k})$, and so Kurt will be provable iff $(b) - \text{Bew}(b, \underline{m})$ is provable.

(1) *If Q is consistent, then Kurt is unprovable.*

Subproof of (1). Suppose Kurt is provable; then there is a proof of Kurt whose GN we will call i. Thus by the properties of Bew, we can prove in Q that $\text{Bew}(\underline{i}, \underline{m})$ and so by E schema we can prove $(Eb)\text{Bew}(b, \underline{m})$.

But we noted above that if Kurt is provable, then so is $(b) - \text{Bew}(b, \underline{m})$ and thus Q is inconsistent.

(2) *If Q is omega-consistent, then Kurt's negation is unprovable.*

Subproof of (2). By part (1), and the fact that omega-consistency entails consistency, Kurt is unprovable and so no number is the number of a proof, i.e., $\cdot - \text{Bew}(0, m)$, $\cdot - \text{Bew}(1, m)$, ... Since Bew is n.r. the system the corresponding formulas are all provable, but Kurt's negation is provable iff $-(b) - \text{Bew}(b, \underline{m})$ is provable and hence the system would be omega-inconsistent if Kurt's negation were provable.

ROSSER'S EXTENSION OF GÖDEL'S THEOREM. *If Q is consistent, then there is a formula which is neither provable nor refutable in Q.*

Strategy. Rosser managed to weaken the hypothesis from omega-consistency to simple consistency by using a more complicated formula whose Gödel number is p and which says that formula p is provable only if there is a shorter proof of its negation. Thus the formula in effect says: 'I am unprovable or the system is inconsistent'. An argument similar to Gödel's shows that p is unprovable if

Q is consistent, but since the negation of p entails consistency, the negation is also unprovable.

Proof. Consider the formula $(b)-\mathrm{Bew}(b, \mathrm{Sub}(29\cdot 31^a, 47, a)) \vee (Ez < b)\mathrm{Bew}(z, 2^{13}{*}\mathrm{Sub}(29\cdot 31^a, 47, a))$. Let its GN be i and consider the formula which is the result of substituting the numeral $\underline{i}$ for the variable a in the formula number i, i.e., $\mathrm{Sub}(29\cdot 31^i, 47, \underline{i})$. This will be the formula $(b)-\mathrm{Bew}(b, \mathrm{Sub}(29\cdot 31^i, 47, \underline{i}) \vee (Ez < b)\mathrm{Bew}(z, 2^{13}{*}\mathrm{Sub}(29\cdot 31^i, 47, \underline{i}))$. Let this formula's GN be p and call it Paul for brevity. We show in two lemmas that Paul is neither provable nor refutable if Q is consistent.

(1) *Suppose Q is consistent and Paul is provable. Then for some n,* $\mathrm{Bew}(n, p)$; *furthermore, by the n.r. representability of* Bew *and* Sub *we know that* $\mathrm{Bew}(\underline{n}, \underline{p})$ *and* $\underline{p} = \mathrm{Sub}(29\cdot 31^i, 47, \underline{i})$ *are provable in Q, hence* $\mathrm{Bew}(\underline{n}, \mathrm{Sub}(29\cdot 31^i, 47, \underline{i})$ *is provable. We also know by the consistency of Q and the n.r. of* Bew *that* $-\mathrm{Bew}(0\cdot 2^{13}{*}\underline{p}), \ldots, -\mathrm{Bew}(\underline{n}-\underline{1}, 2^{13}{*}\underline{p})$ *are all provable. By* 4 *above, we can prove* $(Eb)\cdot\mathrm{Bew}(b, \mathrm{Sub}(29\cdot 31^i, 47, i) \wedge (z < b)-\mathrm{Bew}(z, 2^{13}{*}\mathrm{Sub}(29\cdot 31^i, 47, \underline{i}))$. *But this is Paul's negation and thus Q would be inconsistent; therefore if Q is consistent Paul must be unprovable.*

(2) *Suppose Q is consistent and Paul's negation is provable*; *then let n be the number of such a proof and we can show that* $\mathrm{Bew}(\underline{n}, 2^{13}{*}\underline{p})$ *is provable in Q. Since Q is consistent no number less than n is a proof of Paul, and thus by n.r. of* Bew, $-\mathrm{Bew}(0, \underline{p}) \ldots -\mathrm{Bew}(\underline{n}-\underline{1}, \underline{p})$ *are all provable in Q. Thus by 4 again, we can show that* $(b)[-\mathrm{Bew}(b, \underline{p}) \vee b > \underline{n}]$. *But from this and* $\mathrm{Bew}(n, 2^{13}{*}p)$ *and the provability of* $\underline{p} = \mathrm{Sub}(29\cdot 31^i, 47, \underline{i}))$ *we can find a proof in Q of* $(b)\cdot -\mathrm{Bew}(b, \mathrm{Sub}(29\cdot 31^i, 47, \underline{i})) \vee (Ez < b)\mathrm{Bew}(z, 2^{13}{*}\mathrm{Sub}(29\cdot 31^i, 47, \underline{i}))$. *But this is Paul himself, and thus again Q would be inconsistent if Paul's negation were provable.*

THEOREM. *Q is undecidable.*

Proof. Suppose there were an effective decision procedure for provability in Q, then there would be a predicate $A(a)$ in Q such that $\vdash A(\underline{n})$ iff n is a theorem of Q, and $\vdash -A(\underline{n})$ iff n is a non-theorem of

Q. Consider the formula $-A(\mathrm{Sub}(29\cdot 31^a, 47, a))$; let its Gödel number be j, and consider the formula which results from j by substituting $\underline{j}$ for the variable a in j. This will be the formula $\mathrm{Sub}(29\cdot 31 j, 47, j)$, i.e., $-A(\mathrm{Sub}(29\cdot 31^j, 47, \underline{j}))$. We will now show that if either this formula or its negation is provable then both are.

(1) If $-A(\mathrm{Sub}(29\cdot 31^j, 47, \underline{j}))$ is provable, then by the hypothesis that $A(a)$ n.r. theoremhood, $A(\underline{h})$ will be provable where $\underline{h} = \mathrm{Sub}(29\cdot 31^i, 47, \underline{j})$; but then both $A(\underline{h})$ and $-A(\underline{h})$ are theorems and Q is inconsistent.

(2) Suppose the formula in question is not provable, then by the fact that $A(a)$ n.r. theoremhood $-A(\underline{h})$ will be provable, and thus $-A(\mathrm{Sub}(29\cdot 31^i, 47, \underline{j})$ will be provable contrary to our assumption two lines above that it was not.

Comment. We know that 'n is a theorem' is weakly n.r. by the predicate $(Eb)\mathrm{Bew}(b, \underline{n})$, and the proof just given leaves open the possibility that 'n is a non-theorem' is also weakly n.r. We will now show two further theorems which have as a corollary that if a predicate and its negation are both weakly n.r., then they are n.r., thus showing that non-theoremhood is not even weakly n.r.

THEOREM. *Every predicate n.r. in an effective extension of Q is n.r. in Q. This theorem is partial justification for our choice of n.r.-ability in Q as an explication of effectiveness, since it shows that the choice of some effective extension of Q would give the same class of predicates.*

Proof. Let T be a consistent effective extension of Q; then there is a predicate $\mathrm{Axiom}_T(x)$ which n.r. the GNs of axioms of T and a relation $\mathrm{ImCon}\ (x_1, \dots x_n)_T$ which n.r. the relation which holds between the premises and consequence of the rules of inference of T. Therefore $\mathrm{Bew}_T(m, n)$ will be definable in Q analogously to the definition of Bew. Let $A(a)$ be a formula of T which numeralwise represents some predicate P in T. That is, $P(n)$ iff $\vdash_T A(\underline{n})$ and $-P(n)$ iff $\vdash_T - A(\underline{n})$. We establish that if k is the GN of the formula $A(a)$, then the formula $(Eb)\cdot\mathrm{Bew}_T(b, \mathrm{Sub}(29\cdot 31^{\underline{n}}, 47, \underline{k}))$ & $(c < b) - \mathrm{Bew}_T(c, 2^{13} * \mathrm{Sub}(29\cdot 31^{\underline{n}}, 47, \underline{k}))$ n.r. $P(n)$ in Q. (For convenience we will abbreviate $\mathrm{Sub}(29\cdot 31^{\underline{n}}, 47, \underline{k})$ as '$A(\underline{n})$'.)

(1) If $P(n)$, then the above formula is provable in Q. By the hypothesis that A n.r. P in T, there is a proof of $A\underline{n}$ in T, and thus the first conjunct is easily established; by the consistency of T and an argument analogous to that used in the Rosser argument, the second conjunct will be demonstrable given a specific number of a proof of $A\underline{n}$ in T. On the other hand, if the formula in question is provable, by the fact that Bew_T n.r. provability in T and the fact that A n.r. P in T, $P(n)$ must be true.

(2) $-P(n)$ iff the negation of the above formula is provable in Q. If $(b)(-\mathrm{Bew}_T(b, \text{`}A\underline{n}\text{'}) \vee (Ec < b)\mathrm{Bew}_T(d, \text{`}-A(\underline{n})\text{'})$ is provable in Q, then so is $(b) - \mathrm{Bew}_T(b, \text{`}A(\underline{n})\text{'}) \vee ((Eb)\mathrm{Bew}_T(b, \text{`}-A(\underline{n})\text{'})$. If there is no proof of $A(\underline{n})$ in T, then it must be the case that $-Pn$; if there is a proof of '$A(\underline{n})$' in T, then $-Pn$. (Note that to infer the existence of a proof of the formula from the provability of the existentially quantified sentence in Q assumes the omega-consistency of Q.) If $-Pn$, then by the hypotheses, there is a proof n in T of $-A\underline{n}$, so $\mathrm{Bew}_T(\underline{h}, \text{`}-A(\underline{n})\text{'})$ is provable in Q and by an argument parallel to that for the Rosser theorem, we conclude that the sentence in question is provable.

NORMAL FORM THEOREM. *There is an effective predicate* $T(i, n, b)$ *such that if* $P(n)$ *is any effective predicate, there is an* i *such that for all* n, $P(n) \leftrightarrow (Eb)T(i, n, b)$. The predicate in question is the following: 'b is a proof in Q of the result of substituting the numeral $\underline{n}$ for the variable a in the formula whose GN is i'. By choosing as our i the GN of the formula which n.r. P in Q, we can verify that the predicate has the appropriate properties. To show that it is effective we note that $\mathrm{Bew}(b, \mathrm{Sub}(29 \cdot 31^{\underline{n}}, 47, \underline{i}))$ will n.r. the predicate. We are now in a position to parallel Kleene's proof that $(Ex)T(n, n, x)$ is undecidable.

Note that this T predicate is not exactly the one discussed earlier. That predicate concerned computable functions and this one concerns decidable predicates. Of course a decidable predicate corresponds to a computable function onto zero and one.

THEOREM. (x)-$T(n, n, x)$ *is undecidable.*

Proof. If there were an effective predicate $P(n)$ such that $P(n)$ iff $(x)-T(n, n, x)$, then by the normal form theorem there is an i such that $P(n) \leftrightarrow (Eb)T(n, n, b)$, i.e., $(b)-T(n, n, b) \leftrightarrow (Eb)T(n, n, b)$.

COROLLARY. *$(Eb)T(n, n, b)$ is undecidable.*

COROLLARY. *Any effective, omega-consistent extension of Q is undecidable.*

Proof. $(Eb)T(\underline{n}, \underline{n}, b)$ is provable in Q whenever $(Eb)T(n, n, b)$. If a theory is an omega-consistent extension of Q, then $(Eb)T(\underline{n}, \underline{n}, b)$ will be provable only if $(Eb)T(n, n, b)$ is true. Thus if such an extension of Q were decidable, we could decide for any n whether $(Eb)T(n, n, b)$, but this is impossible.

COROLLARY. *The set of truths expressible in the vocabulary of Q is undecidable.*

Proof. The set of truths expressible in the vocabulary of Q is an omega-consistent extension of Q; if it were decidable then we could take every truth as an axiom and we would have an effective, omega-consistent extension which is decidable.

EXERCISE 2. Show that if P and $-P$ are both weakly representable, then P is n.r. (*Hint*: Let $A(a)$ be the predicate which weakly n.r. P and $B(a)$ the predicate which weakly n.r. $-P$, and use the formula $(Eb)\text{Bew}(b, `A\underline{n}) \ \& \ (c < b) - \text{Bew}(c, `B\underline{n}')$.)

EXERCISE 3. Show that $(b)-T(n, n, b)$ is not weakly n.r.

THEOREM. *Every consistent effective extension of Q is undecidable.*

Proof. By the generalized Rosser theorem, any effective extension of Q is incomplete. We prove now a lemma which asserts that any effective decidable theory can be extended to a complete theory. Therefore if there were a decidable extension of Q it could be extended to a complete extension of Q.

LEMMA. *If T is effective and decidable then there is an effective complete extension of T.*

Proof. Let Bew_T be the proof predicate for T, we form T' by imitating the maximal consistent set construction for T and making all added formulas axioms. We can consider formulas in order of increasing GNs and add all those which do not permit the derivation of a specified contradiction. This process is effective because T is decidable and hence any finite extension of T is also.

COROLLARY. *Q is essentially undecidable.*

DEFINITION. A theory is essentially undecidable if there is no decidable extension.

CHURCH'S THEOREM. *Pure first-order quantification theory is undecidable.*

Proof. By the theorem on the eliminability of function symbols there is a theory Q^* without function symbols such that $Q^* \vdash A^*$ iff $Q \vdash A$. By the theorem on p. 47, there is a theory without identity Q_F such that $Q_F \vdash A_F$ iff $Q^* \vdash A^*$. Thus by the deduction theorem we know that $Q \vdash A$ iff $\vdash Q_F \supset A_F$. Since problems of the first form are undecidable so are those of the second. By construction $Q_F \supset A_F$ contains no function symbols and does not contain the identity sign.

THEOREM. *If Q is consistent, then there are consistent but omega-inconsistent theories.*

Proof. Kurt's negation is consistent with Q if Q is consistent since otherwise by double negation elimination Kurt would be provable and Q would be inconsistent. Therefore adding Kurt's negation to Q gives a consistent but omega-inconsistent system (which could be Q).

THEOREM. *If Q is consistent it has non-standard models.*

Proof. If Q is consistent then the omega-inconsistent extension just constructed has a model which is necessarily non-standard since no object assigned to a numeral satisfies the formula $(Eb)\cdot\mathrm{Bew}(b, \underline{m})$.

We proved a more general form of this theorem earlier but that proof used the compactness theorem and this one depends only on the completeness theorem. Also, in this theorem we can specify a particular formula in our number theoretic vocabulary which is satisfied in the non-standard model though intuitively false.

CHAPTER VIII

GÖDEL'S SECOND INCOMPLETENESS THEOREM

One of the main goals of research in the foundations of mathematics in the 1920's was to find a consistency proof for number theory. Not that the consistency of number theory was considered to be very dubious, but the consistency of set theory was considered a partially open question and finding a convincing consistency proof for number theory seemed a natural first step in that direction. The possibility of such a proof does not seem too unlikely since we can formalize the statement of consistency in a rather simple form: $(x) - \text{Bew}(x, \text{`}0 = 1\text{'})$. It is true that one would have to use some number theoretic principles or their equivalents to prove a statement of this form, but there was reason to hope that the proof could be carried out in a relatively 'small' subsystem, for example, using induction only for quantifier free formulas. Gödel showed, however, in 1931 that no such proof was possible.

The idea of Gödel's proof is simple. The proof of the first Gödel incompleteness theorem shows that if number theory is consistent then a particular sentence is unprovable. If we could show that the reasoning of that argument could be formalized in the system then we would be able to show that

$$\vdash (x) - \text{Bew}(x, \text{`}0 = 1\text{'}) \supset (x) - \text{Bew}(x, \underline{k}),$$

where k is the number of the Gödel sentence in the first theorem. Since we know that the consequent of this conditional is not provable in N if N is consistent, we could show that if N is consistent then $(x) - \text{Bew}(x, \text{`}0 = 1\text{'})$ is not provable.

Stating a general form of the Second Incompleteness Theorem is more difficult than stating a general form of the First Incompleteness Theorem because more attention must be paid to the formula which is used to represent the proof relation. In the First Theorem we only needed to know that the two-place relation numeralwise represents the relation '$B_1, \ldots B_k$ is a proof of A', but for the Second Incompleteness Theorem further conditions are required in order to ensure that the argument of the First Theorem is formalizable. Our procedure will be to review the argument for the First Theorem and then to state some general conditions which suffice to formalize the argument. We will then show that N meets these conditions. Later we will give some examples of proof relations which do not satisfy these conditions and give an alternative set of conditions.

In proving the first theorem we first constructed a sentence with Gödel number k which was of the form $(b)-\text{Bew}(b, t)$ where $\vdash t = \underline{k}$. Next we showed that if this sentence was provable then N was inconsistent by the following argument:

(i) If $(b)-\text{Bew}(b, t)$ is provable then there is a number of that proof, call it m, such that $\vdash \text{Bew}(\underline{m}, \underline{k})$

(ii) If $\vdash \text{Bew}(\underline{m}, \underline{k})$ then $\vdash -(b)-\text{Bew}(b, \underline{k})$

(iii) If $\vdash -(b)-\text{Bew}(b, \underline{k})$ then $\vdash -(b)-\text{Bew}(b, \underline{t})$

(iv) If $\vdash (b)-\text{Bew}(b, \underline{t})$ and $\vdash -(b)-\text{Bew}(b, \underline{t})$, then $\vdash 0 = 1$.

To formalize the proof in the system itself we would need to show:

(i) $\vdash (Eb)\text{Bew}(b, \underline{k}) \supset (Ec)(Ed)\text{Bew}(c, \text{'Bew}(\underline{d}, \underline{k})\text{'})$

(ii) $\vdash (Ec)(Ed)\text{Bew}(c, \text{'Bew}(\underline{d}, \underline{k})\text{'}) \supset (Ec)\text{Bew}(c, \text{'}-(b)-\text{Bew}(b, \underline{k})\text{'})$

(iii) $\vdash (Ec)\text{Bew}(c, \text{'}-(b)-\text{Bew}(b, \underline{k})\text{'})$

$\supset (Ec)\text{Bew}(c, \text{'}-(b)-\text{Bew}(b, t)\text{'})$

(iv) $\vdash (Ec)\text{Bew}(c, \underline{k}) \supset [(Ec)\text{Bew}(c, 2^{13} * \underline{k}) \supset (Ec)\text{Bew}(c, \text{'}0 = 1\text{'})]$.

from which it would follow that

$\vdash (Eb)(\text{Bew}(b, \underline{k}) \supset (Ec)\text{Bew}(c, \text{'}0 = 1\text{'})$, and by contraposition we would obtain the result we need.

In order to carry out this formalization it will suffice to have the following properties of Bew:

(A) If $\vdash A$, then $\vdash (Eb)\text{Bew}(b, \text{'A'})$

(B) $\vdash (Eb)\text{Bew}(b, \text{'}A \supset B\text{'})$

$\supset [(Eb)\text{Bew}(b, \text{'}A\text{'}) \supset (Eb)\text{Bew}(b, \text{'}B\text{'})]$

(C) $\vdash (c)(b)[\text{Bew}(b, c) \supset (Ed)\text{Bew}(d, \text{'Bew}(\underline{b}, \underline{c})\text{'})]$.

(i) follows from (C) by quantification theory. (ii) follows from the fact that $\vdash \text{Bew}(\underline{d}, \underline{k}) \supset -(b)-\text{Bew}(b, \underline{k})$ by (A) and (B). (iii) follows from the fact that $\vdash \underline{k} = t \supset [-(b)-\text{Bew}(b, \underline{k}) \supset -(b)-\text{Bew}(b, t)]$ by (A) and (B) together with the application of (A) to $\vdash k = t$. Finally, (iv) follows from $\vdash A \supset (-A \supset 0 = 1)$ by (A) and two applications of (B).

EXERCISE 1. Write out the details of the proof that (i)–(iv) hold if (A)–(C) do.

HILBERT-BERNAYS GENERALIZATION OF GÖDEL'S SECOND INCOMPLETENESS THEOREM. *If a system of number theory T has a function expression which numeralwise represents the substitution operation and a proof predicate* Bew_T *which satisfies conditions* (A)–(C), *then if the system is consistent* $(b) - \mathrm{Bew}_T(b, \text{`}0 = 1\text{'})$ *is not a theorem.*

Proof. By the previous exercise, if (A)–(C) hold then

$$\vdash (b) - \mathrm{Bew}_T(b, \text{`}0 = 1\text{'}) \supset (b) - \mathrm{Bew}_T(b, \underline{k})$$

and if substitution is numeralwise represented then by the first incompleteness theorem we know that if the system is consistent $(b) - \mathrm{Bew}_T(b, \underline{k})$ is not a theorem. Thus if the system is consistent $(b) - \mathrm{Bew}_T(b, \text{`}0 = 1\text{'})$ is also not a theorem.

GÖDEL'S SECOND INCOMPLETENESS THEOREM FOR *N*. *If N is consistent,* $(b) - \mathrm{Bew}(b, \text{`}0 = 1\text{'})$ *is not provable.*

Proof. We will show that the proof predicate we defined in the previous chapter for *N* satisfies (A)–(C). A has already been established in the previous chapter, so we may proceed with B. If we are given a proof of $A \supset B$ and a proof of *A* we can obtain a proof of *B* by putting one of the proofs after the other and appending *B* as a last line – the difficulty is to show that this argument can be carried out formally in *N*.

EXERCISE 2. Define a functional expression $Cd(a, b)$ of *N* such that *Cd* numeralwise represents the function such that $Cd(\text{`}A\text{'}, \text{`}B\text{'}) = \text{`}(A \supset B)\text{'}$.

We will now define a functional expression which will give us the proof number if *c* is a proof of $Cd(a, b)$ and *d* is a proof of *a*:

$f(b, c, d) = c*d*2^b$. What we must prove is first that:

$$\text{Bew}(c, Cd(a, b)), \text{Bew}(d, a) \vdash \text{Bew}(f(a, b, c, d), b).$$

It is trivial that

$$\text{Bew}(c, Cd(a, b)), \text{Bew}(d, a) \vdash \text{ImCon}((c)_{\text{gp}(c)}, (d)_{\text{gp}(d)}, b),$$

and so from the same assumptions

$$\vdash \text{ImCon}(f(a, b, c, d)_{\text{lh}(c)}, f(a, b, c, d)_{\text{lh}(c*d)}, f(a, b, c, d)_{\text{lh}(f(a, b, c, d))}).$$

We also know that

$$\text{Bew}(c, Cd(a, b)) \vdash (i \leqslant \text{lh}(c))[(Ax((f(a, b, c, d))_i) \vee (Ej, k < i) \text{ ImCon}((f(a, b, c, d))_j, (f(a, b, c, d))_k, (f(a, b, c, d))_i)].$$

Furthermore,

$$\text{Bew}(d, a) \vdash (i \leqslant \text{lh}(c*d))[i > \text{lh}(c) \supset (Ax((f(a, b, c, d))_i) \vee (Ej, k < i) \text{ ImCon}((f(a, b, c, d))_j, (f(a, b, c, d))_k, (f(a, b, c, d))_i)].$$

Putting together these last three deductions from assumptions we obtain

$$\text{Bew}(c, Cd(a, b)), \text{Bew}(d, a) \vdash (i \leqslant \text{lh}(f(a, b, c, d)))[Ax(f(a, b, c, d)_i \vee (Ej, k < i) \text{ ImCon}((f(a, b, c, d))_j, (f(a, b, c, d))_k, (f(a, b, c, d))_i)],$$

and so by the definition of Bew,

$$\text{Bew}(c, Cd(a, b)), \text{Bew}(d, a) \vdash \text{Bew}(f(a, b, c, d), b).$$

From which we can infer

$$\vdash (a)(b)[(Ec)\text{Bew}(c, Cd(a, b)) \supset ((Ed)\text{Bew}(d, a) \supset (Ed)\text{Bew}(d, b))],$$

which was what was to be established.

In order to prove that N meets condition C, we must consider how we would prove Bew(m, n) in any particular case. Such a proof must consist of showing for each $i \leqslant \mathrm{lh}(m)$ that $(m)_i$ is an axiom or is an immediate consequence of earlier lines, and also showing of the values of i for which this is proved that they *are* all the values of i less than or equal to lh(m). In order to show that we can prove Bew(m, n) we must then analyze these subproofs and so on. Instead of pursuing this route to a proof we will establish a more general theorem.

We will begin by showing that the simplest equations involving successor, addition, multiplication, etcetera are provably represented and then follow the development of the proof that Bew is representable.

If $x = y$, then $\underline{x} = \underline{y}$ is an instance of Axiom 6 so

$$\vdash (a)(a = a \supset \underline{x} = \underline{y})$$
$$\vdash (a)(a = a)$$
$$\vdash \underline{x} = \underline{y}$$

will be a proof and we can show that this is provable. Thus,

$$\vdash x = y \supset (Ez)\mathrm{Bew}(z, `\underline{x} = \underline{y}').$$

If $x = 0$ and $y \neq x$, then similarly using Axioms 5 and 8, and 10

$$\vdash y \neq 0 \supset (Ez)\mathrm{Bew}(z, `y \neq 0'),$$

and using Axioms 5, 7, 8 and 10

$$\vdash 0 \neq x \supset (Ez)\mathrm{Bew}(z, `0 \neq \underline{x}').$$

We note next that $\vdash (y)(x)(y \neq x \leftrightarrow y' \neq x')$, thus invoking conditions (A) and (B),

$$y' \neq x', y \neq x \supset (Ez)\mathrm{Bew}(z, `\underline{x} \neq \underline{y}') \vdash (Ez)\mathrm{Bew}(z, `\underline{x}' \neq \underline{y}'),$$

and so

$$\vdash (y = x \supset (Ez)\mathrm{Bew}(z, `\underline{x} \neq \underline{y}')) \supset (x' = y' \supset (Ez)\mathrm{Bew}(z, `\underline{x}' \neq \underline{y}')).$$

This formula together with our previous theorems provide the premises for two applications of induction in N which yield the conclusion $\vdash (x)(y)[x \neq y \supset (Ez)\text{Bew}(z, `\underline{x} \neq \underline{y}')]$.

If $x + 0 = z$, then the formula will be provable by using Axiom schema 5 to instantiate Axiom 11, and again it is trivial to show $\vdash x + 0 = z \supset (Ew)\text{Bew}(w, `\underline{x} + 0 = \underline{z}')$. Furthermore, we can show that $\vdash x + y' = z \leftrightarrow (Ev)(x + y = v \wedge v' = z)$. Thus using conditions (A) and (B) and the fact that $\vdash v' = z \supset (Ew)\text{Bew}(w, `\underline{v}' = \underline{z}')$, $x + y = v \supset (Ew)\text{Bew}(w, `\underline{x} + \underline{y} = \underline{v}')$, $x + y' = z \vdash (Ew)\text{Bew}(w, `\underline{x} + \underline{y}' = \underline{z}')$. Using the deduction theorem twice we find that we have the antecedents of an instance of the induction axiom in N and can infer $(x)(y)(z)(x + y = z \supset (Ew)\text{Bew}(w, `\underline{x} + \underline{y} = \underline{z}')$.

It is fairly easy to show that $\vdash x + y \neq z \leftrightarrow (Ev)(x + y = v \wedge v \neq z)$. From this fact together with the theorems just proved and conditions (A) and (B), it follows that $\vdash (x)(y)(z)(x + y \neq z \supset (Ew)\text{Bew}(w, `\underline{x} + \underline{y} \neq \underline{z}'))$. Since we will frequently need to make reference to such conditions, we will introduce some special terminology. If a predicate $Ax_1 \ldots x_n$ is numeralwise represented by $Bx_1 \ldots x_n$ and $\vdash_N Bx_1 \ldots x_n \supset (Ew)\text{Bew}(w, `B\underline{x}_1 \ldots \underline{x}_n')$ and $\vdash_N -Bx_1 \ldots x_n \supset (Ew)\text{Bew}(w, `-B\underline{x}_1 \ldots \underline{x}_n')$ we will say that A *is provably represented* in N. (We will abbreviate this as 'PR in N' and when it is clear that N is meant we will simply say PR.) We can summarize our results thus far as

LEMMA 1. *Predicates formed out of identity, successor, addition and negation are* PR.

EXERCISE 3. Prove that $xy = z$ is PR in N.

EXERCISE 4. Prove that $x^y = z$ is PR in N.

EXERCISE 5. Prove that predicates formed by truth functional combination of PR predicates are PR.

Our next objective is to prove that $x < y$ is PR. The positive part is easy since by Lemma 1, $\vdash w + x' = y \supset (Ez)\text{Bew}(z, `\underline{w} + \underline{x}' = \underline{y}')$ and by quantification theory and conditions (A) and (B) $\vdash (Ez)\text{Bew}(z, `\underline{w} +$

$\underline{x}' = \underline{y}') \supset (Ez)\text{Bew}(z, \text{`}(Ew)\underline{w} + \underline{x}' = \underline{y}\text{'})$, so by quantification theory $\vdash (Ew)(\underline{w} + \underline{x}' = \underline{y}) \supset (E\underline{z})\text{Bew}(z, \text{`}(Ew)\underline{w} + \underline{x}' = \underline{y}\text{'})$. In order to prove the negative portion it will be useful to establish that $\vdash_N x < y' \leftrightarrow x < y \vee x = y$, which is by definition $\vdash_N (Ew)(w + x' = y') \leftrightarrow (Ev)(v + x' = y \vee x = y)$. In order to establish this it will first be useful to establish $\vdash_N (x)(0 + x = x)$ and $\vdash_N (x)(y)(x' + y = x + y')$. Both of these require the use of induction in N.

Clearly by Axiom 11, $\vdash 0 + 0 = 0$. Furthermore, $\vdash 0 + x' = (0 + x)'$ so $0 + x = x \vdash 0 + x' = x'$. So by induction $\vdash (x)(0 + x = x)$. Using this theorem we show that $\vdash x' + 0 = 0 + x'$, and by Axiom 12, $\vdash 0 + x' = (0 + x)'$, so using our theorem again together with transitivity of identity $\vdash x' + 0 = x + 0'$. By Axiom 12 again, $\vdash x' + y' = (x' + y)'$. Thus $x' + y = x + y' \vdash x' + y' = (x + y')'$, and so by Axiom 12 and the deduction theorem $\vdash (y)((x)(x' + y = x + y') \supset (x)(x' + y' = x + y''))$. Finally, by induction in $N \vdash (y)(x)x' + y = x + y'$.

Returning to our main proof we see that by Axioms 10 and 12, $w + m' = n' \vdash (w + m)' = n' \wedge (Ev)(w = v' \vee w = 0)$. Distributing $\vee$ over $\wedge$, $w + m' = n' \vdash (0 + m)' = n' \vee (Ev)(v' + m)' = n'$. Using the theorems just proved $w + m' = n' \vdash (m + 0)' = n' \vee (Ev)(v + m')' = n'$. Finally, by Axioms 9, 11 and 12 $w + m' = n' \vdash m' = n' \vee (Ev)(v + m' = n)$, and so by quantification theory and definitions $\vdash m < n' \leftrightarrow m < n \vee m = n$.

EXERCISE 6. Prove $\vdash (x < y \vee x = y) \rightarrow x < y'$.

It is trivial to show that $\vdash (x)x \not< 0$, and so it is not difficult to show that $\vdash (x)[x \not< 0 \supset (Ez)\text{Bew}(z, \text{`}\underline{x} \not< 0\text{'})$. By our previous theorem, $x \not< y' \leftrightarrow x \not< y \wedge x \neq y$. Thus by conditions (A) and (B), and the fact that $=$ is PR $x \not< y \supset (Ez)\text{Bew}(z, \text{`}x \not< y\text{'})$, $x \not< y' \vdash (Ez)\text{Bew}(z, \text{`}\underline{x} \not< \underline{y}'\text{'})$. So by induction in $N$, $\vdash (x)(y)(x \not< y \supset (Ez)\text{Bew}(z, \text{`}x \not< y\text{'}))$.

Next we wish to establish the correspondingly stronger forms of (5a), (b), (c) on page 57. That is, we wish to show that if $A(x, y)$ is provably represented in N then so are $(Ex < w)A(\underline{x}, \underline{y})$, $(x < w)A(x, y)$ and $z = \mu x(x < w \wedge A(x, y))$. We will assume that B provably represents A. Since $\vdash (x)x \not< 0$, it is obvious that

$$\vdash (Ex)(x < 0 \wedge B(x, y)) \supset (Ez)\text{Bew}(z, \text{`}(Ex)(x < 0 \wedge B(\underline{x}, \underline{y})\text{'})$$

and

$$\vdash -(Ex)(x < 0 \wedge B(x, y) \supset (Ez)\text{Bew}(z, `-(Ex)(\underline{x} < 0 \wedge B(\underline{x}, \underline{y}))').$$

In order to show that the induction step for the positive case is provable in N, we note that

$$\vdash (Ex)(x < w' \wedge B(x, y)) \supset (Ex)(x < w \wedge B(x, y) \vee B(w', y).$$

Using conditions (A) and (B), the assumption that B is PR, the fact that $<$ is PR, and the fact that truth functions preserve PR ability,

$$\vdash (Ex)(x < w \wedge B(x, y)) \supset (Ez)\text{Bew}(z, `(Ex)(\underline{x} < \underline{w} \wedge B(\underline{x}, \underline{y}))')$$

$$(Ex)(x < w' \wedge B(x, y)) \supset (Ez)\text{Bew}(z, `(Ex)(x < \underline{w}' \wedge B(\underline{x}, \underline{y}))').$$

Thus by induction in N it follows that

$$\vdash (w)[(Ex)(x < w \wedge B(x, y)) \supset (Ez)\text{Bew}(z, `(Ex)(\underline{x} < \underline{w} \wedge B(x, \underline{y}))')].$$

To prove the negative half of the PR condition is satisfied, we use the biconditional at the beginning of the previous paragraph, negate both sides and apply some quantificational logic,

$$\vdash [-B(w', y) \wedge -(Ex)(x < w, B(x, y)] \supset -(Ex)(\underline{x} < w' \wedge B(\underline{x}, \underline{y}).$$

Again invoking conditions (A) and (B), and facts about B, $<$ and truth functions

$$-(Ex)(x < w \wedge B(x, y)) \supset (Ez)\text{Bew}(z, `-(Ex)(\underline{x} < \underline{w} \wedge B(\underline{x}, \underline{y}))')$$

$$\supset -(Ex)(x < w' \wedge B(x, y)) \supset (\text{Ez})\text{Bew}(z, `-(Ex)(\underline{x} < \underline{w}' \wedge B(\underline{x}, \underline{y}))').$$

EXERCISE 7. Give the details of the proof that $(Ex < w)A(x, y)$ is PR.

EXERCISE 8. Prove that $(x < w)A(x, y)$ is PR, if $A(x, y)$ is.

Just as in the previous chapter proving (5a) and (b) simplified the remaining arguments, we have now done the hardest part of the work. If $A(x, y)$ is PR, then $z = \mu x[x < w \wedge A(x, y)]$ is PR since $<$ is PR and bounded quantification and truth functional composition preserve PR ability.

LEMMA. Bew *meets condition* (C).

Proof. We can now follow the series of definitions leading from (5) on page 57 to the definition of Bew and check that each definition combines predicates which were shown previously to be PR in ways which preserve provable representability. Thus Bew is PR. This completes our proof that N satisfies the conditions (A)–(C) and thus that Gödel's Second Incompleteness Theorem applies to N.

Checking our proof of the Second Incompleteness Theorem for N, it is not difficult to see that the crucial point was that Bew is a formula which represents the proof relation in a suitable way. Some conditions are necessary for the Second Incompleteness Theorem in that using certain unnatural proof predicates a statement which looks like a consistency assertion is provable. For example, suppose we define $B_1(x, y)$ as $\mathrm{Bew}(x, y) \wedge (z \leqslant x) - \mathrm{Bew}(z, \text{‘}0 = 1\text{’})$. Assuming N is consistent, $B_1(x, y)$ numeralwise represents the proof relation in N, because if $\mathrm{Bew}(m, n)$ then $\vdash \mathrm{Bew}(m, n)$ and if N is consistent $\vdash - \mathrm{Bew}(0, \text{‘}0 = 1\text{’})$, $\vdash - \mathrm{Bew}(1, \text{‘}0 = 1\text{’})$ and so $B_1(m, n)$. But notice that $(x) - B_1(x, \text{‘}0 = 1\text{’})$ by definition is equivalent to $(x) - [\mathrm{Bew}(x, \text{‘}0 = 1\text{’}) \wedge (z \leqslant x) - \mathrm{Bew}(z, \text{‘}0 = 1\text{’})]$, which is provable since $-[\mathrm{Bew}(x, \text{‘}0 = 1\text{’}) \wedge - \mathrm{Bew}(x, \text{‘}0 = 1\text{’})]$ is provable.

Thus if N is consistent, B_1 numeralwise represents the proof relation and yet $(x) - B_1(x, \text{‘}0 = 1\text{’})$ is a theorem. The unnaturalness in this proof predicate B_1 arose from the fact that the requirement of consistency was ‘built into’ the definition of a proof. A similarly peculiar proof predicate can be found by building in the consistency requirement in the specification of the axioms. We can readily define a three-place relation $B(n, x, y)$ which represents ‘x is a proof of y and x uses only axioms with Gödel numbers less than n’ as

$\text{Bew}(x, y) \wedge (i < \text{lh}(x))[Ax((x)_i) \supset (x)_i < n]$. We can now define a two-place proof predicate $B_2(x, y)$ as $B_2(x, y) =_{\text{df}} B((x)_1, (x)_2, y) \wedge (z) - B((x)_1, z, \text{`}0 = 1\text{'})$. In other words, $(Ez)B_2(z, y)$ will hold iff y is provable in N from a finite set of axioms of N which are consistent. Thus if N is consistent $(Ez)B_2(z, y)$ will coincide with $(Ez)\text{Bew}(z, y)$. Moreover, it can be shown that any finite set of axioms of N can be proved consistent in N, and thus $B_2(x, y)$ is numeralwise representable. Obviously, $\vdash (z) - B_2(z, \text{`}0 = 1\text{'})$.

EXERCISE 9. Which of the conditions (A)–(C) does B_1 fail to meet? B_2?

Thus it is necessary that in order to apply the Second Incompleteness Theorem the specification of both the axioms and of the consequence relation have to be appropriate.

FEFERMAN'S GENERALIZATION OF GÖDEL'S SECOND INCOMPLETENESS THEOREM. *If* $A(x)$ *is a formula of a first order theory* T *(which includes* N*) which provably numeralwise represents axiomhood in* T *and* $\text{IC}(x, y, z)$ *is a formula of* T *which provably numeralwise represents immediate consequence and if* $B(x, y)$ *is defined as* $(i \leq \text{lh}(x))[A((x)_i \vee (Ez, w < i)(\text{IC}((x)_z, (x)_w, (x)_i)]$, *and if* $\text{Con}(T)$ *is defined as* $(x) - B(x, \text{`}0 = 1\text{'})$, *then if* T *is consistent* $\text{Con}(T)$ *is not a theorem of* T.

Proof. Since A numeralwise represents axiomhood and IC the consequence relation and since the definition of B combines these in ways that preserve numeralwise representability, B will n.r. the proof relation and so condition (A) will be satisfied. Since they provably represent axiomhood and consequence and since the way in which they are combined in the definition of B also preserves provable representability, condition (C) is also satisfied. The proof that condition (B) is satisfied consists of verifying that all of the information needed in defining the function f on page 75 and proving the properties of f are provided in the statement of the theorem.

The Second Incompleteness Theorem is undoubtedly the one result in logic which is most frequently misstated and misused for philosophical purposes. One form of 'argument' is the following: Given any sufficiently rich consistent formal system there is a sentence which is not provable but which we can see to be true, therefore, our capacities exceed those of any formal system. We *can* show that for any sufficiently rich consistent system there is an unprovable sentence which expresses consistency and thus is true. But this can also be proved in *N*. What the argument requires for its conclusion is that for such systems we can show that the sentence is unprovable and we can do that only if we can prove that the system is consistent.

In other cases philosophers have argued that we are not representable by formal systems because we can prove our own consistency. As the past few pages should amply demonstrate, unless such claims are made precise by specifying the exact form of the consistency statement there is no reason to think that the Incompleteness Theorem can be invoked to show that the corresponding statement is not formally provable. Furthermore, some of the informal arguments used to establish 'our' consistency are demonstrably inconsistent. (For example, the principles in Exercise 10 of this chapter were used by Lucas in 'Minds, Machines and Gödel'.)

We conclude this chapter with another application of the conditions (A)–(C). One of the natural questions which arises from the First Incompleteness Theorem concerns what types of extension we might consider for *N*. We could, of course, add the Gödel sentence itself, or (having seen the Second Incompleteness Theorem) perhaps the consistency statement, but this is not a very general method. One class of formulas which we might consider would be those of the form $(Ez)\mathrm{Bew}(z, `A') \supset A$. If our formal system is correct then all of the statements of this form will be true. By trivial facts about the propositional calculus, if $A$ is a theorem then the corresponding sentence $(Ez)\mathrm{Bew}(z, `A') \supset A$ will be provable. By the Second Incompleteness Theorem and propositional calculus, if $-A$ is a theorem $(Ez)\mathrm{Bew}(z, `A') \supset A$ will be unprovable since if it were $-(Ez)\mathrm{Bew}(z, `A')$ would be provable and then by the conditions (A)–(C), $\vdash -(Ez)\mathrm{Bew}(z, `0 = 1')$ since $\vdash 0 = 1 \supset A$. Thus adding all of the sentences of the form in question would give a significant extension of *N*. In this extension the consistency of *N* would be provable.

We have shown that sentences of the form $(Ez)\text{Bew}(z, `A\text{'}) \supset A$ are provable if $A$ is provable and not provable if $-A$ is provable. This leaves open the question of the status of such sentences when $A$ is undecidable. Löb devised an ingenious argument rather like Gödel's which shows that if $A$ is provable then $(Ez)\text{Bew}(z, `A\text{'}) \supset A$ is also. Intuitively, the argument consists of constructing a sentence K which says 'If sentence K is provable then A'. It is not difficult to show that if K is provable then A is provable, and so if the provability of A entailed A itself then we would have a proof of A. As in the case of Gödel's theorem, of course, we must see that we can construct such a sentence and formalize the reasoning.

LÖB'S THEOREM. *If A is unprovable, then $(Ez)\text{Bew}(z, `A\text{'}) \supset A$ is unprovable.*

Proof. Consider the formula $-(b) - \text{Bew}(b, \text{Sub}(29 \cdot 31^{a}, 47, a)) \supset A$. Let the Gödel number of this formula be j and consider the formula which results from substituting the numeral $\underline{j}$ for the variable a in formula j, i.e., $\text{Sub}(29 \cdot 31^{j}, 47, j)$, which will be the formula $-(b) - \text{Bew}(b, \text{Sub}(29 \cdot 31^{\underline{j}}, 47, \underline{j})) \supset A$. If we let n be the Gödel number of this sentence, we know that the sentence will be provable iff $-(b) - \text{Bew}(b, \underline{n}) \supset A$ is provable. By condition (B) above,

$$\vdash (Ec)\text{Bew}(c, `(Eb)\text{Bew}(b, \underline{n}) \supset A\text{'}) \supset [(Ec)\text{Bew}(c, `(Eb)\text{Bew}(b, \underline{n})\text{'} \supset (Ec)\text{Bew}(c, `A\text{'})] \text{ and since } \vdash \underline{n} = `(Eb)\text{Bew}(b, \underline{n}) \supset A\text{'}$$

$$\text{(I)} \qquad \vdash (Eb)\text{Bew}(b, \underline{n}) \wedge (Ec)\text{Bew}(c, `(Eb)\text{Bew}(b, \underline{n})\text{'}) \supset (Ec)\text{Bew}(c\ `A\text{'}).$$

By condition C, $\vdash \text{Bew}(b, \underline{n}) \supset (Ec)\text{Bew}(c, `\text{Bew}(\underline{b}, \underline{n})\text{'})$, and since $\vdash \text{Bew}(\underline{b}, \underline{n}) \supset (Eb)\text{Bew}(b, \underline{n})$, we can use conditions (A) and (B) to obtain $\vdash (Eb)\text{Bew}(b, \underline{n}) \supset (Ec)\text{Bew}(c, `(Eb)\text{Bew}(b, \underline{n})$. Thus using (I) $\vdash (Eb)\text{Bew}(b, \underline{n}) \supset (Ec)\text{Bew}(c, `A\text{'})$. If $\vdash (Ec)\text{Bew}(c, `A\text{'}) \supset A$, $\vdash (Eb)\text{Bew}(b, \underline{n}) \supset A$. But this is the formula whose Gödel number is n, and so then we would also have $\vdash (Eb)\text{Bew}(b, \underline{n})$ and so $\vdash A$. Thus if $(Ec)\text{Bew}(c, `A\text{'}) \supset A$ is provable, then A is also.

EXERCISE 10 (due to P. Benacerraf). Show that if an extension T of N has a formula P with a single free variable and which is such that (1) whenever $\vdash_T A$, then $\vdash_T P(`A')$ and (2), for all $A$, $\vdash P(`A') \supset A$, then T is inconsistent.

EXERCISE 11 (due to S. Kripke). We will use R for the Rosser sentence, Con N for the sentence $-(Ec)$Bew$(c, `0 = 1')$, $N + R$ for the system obtained by adding R to N as an axiom and Con $(N + R)$ for the corresponding consistency statement. Show why the following argument is fallacious: Rosser's theorem shows that if Con N then neither $\vdash R$ nor $\vdash - R$. Thus if Con N, Con $(N + R)$ and so formalizing this we can show $\vdash_{N+R}$Con $N \supset$ Con $(N + R)$. But since $\vdash_N R \supset$ Con N, $\vdash_{R+N}$Con N, and so $\vdash_{N+R}$Con$(N + R)$. Thus by Gödel's second theorem $N + R$ is inconsistent, so by Rosser's theorem N is inconsistent!

CHAPTER IX

TARSKI'S THEOREMS AND THE DEFINITION OF TRUTH

Our next subject is that of formalizing definitions of truth for first order theories within first order theories. We already have two impossibility theorems, one of which was proved earlier and one of which is an easy consequence of earlier theorems. First, we might consider the possibility of finding for arbitrary theories a way of defining a predicate T such that T would numeralwise represent truth in the theory. That is, we would hope for a general method such that given an effective theory H and an interpretation we could define a predicate T_H such that $\vdash_Q T_H(\underline{n})$ iff n is the Gödel number of a true sentence of H. However, since we know that the set of true sentences of Q is not even recursively enumerable we know that we cannot numeralwise represent truth in Q in any effective extension of Q. The next best sense of definition that we could hope for would be that we could define a predicate T such that for every sentence of H we could prove $T(\text{‘}A\text{’}) \leftrightarrow A$ in some suitable theory. We would, of course, want the theory which defined the predicate T to be consistent. If a theory has as consequences all of the sentences of the form $T(\text{‘}A\text{’}) \leftrightarrow A$ where ‘A’ is a suitable representation of a sentence A of H and the theory is consistent, we shall say that it satisfies *Tarski's condition* for H.

We proved in Chapter VII that there is no theory in the vocabulary of N which satisfies Tarski's condition for N. Our main two theorems of this section, both proved originally by Tarski, will be that for any theory as rich as N no truth definition that satisfies Tarski's condition for a theory can be given in the same vocabulary, but that by extending the vocabulary we can always give a theory which satisfies Tarski's condition.

We must begin by being more explicit about the nature of the representation of the vocabulary of the language for which we are defining truth. We will use OL as a name for the object language, the one for which we are defining truth, and ML as a name for the metalanguage in which we are defining truth. We will say that ML represents the syntax of OL iff there are expressions of ML which meet the following conditions:

There is a mapping from the variables, function letters, predicate letters, terms and formulas of OL to terms of ML such that:

(0) There is an expression Con(x, n) such Con(t, n) is provable iff t represents the nth constant of the language.

(1) There is an expression Var(x, y) such that Var(x, n) is provable iff x represents the nth variable of OL.

(2) For each n such that OL contains n-place function letters f_i^n, there is an $n+1$ place function term FnTerm such that FnTerm $(i, t_1, \ldots t_n) = s$ is provable iff s represents the term formed by applying the ith n place function letter to the terms represented by $t_1, \ldots t_n$.

(3) For each n such that OL contains n-place predicates, there is an $n+1$ place functional expression Pred$n$ such that Pred$n(i, t_1, \ldots t_n) = s$ is provable iff s represents the atomic formula which results from applying the ith n-placed predicate to the terms represented by $t_1, \ldots t_n$.

(4) There is a three-place function expression Subst such that Subst(t, t_2, t) $= s$ is provable iff s represents the formula which results from substituting the term represented by t_1 for the term represented by t_2 in the formula represented by t.

(5) There is a one-place function expression Neg such that Neg(s) $= t$ is provable iff t represents the negation of the formula represented by s.

(6) There is a two-place function expression Imp(s, t) such that Imp(s, t) $= r$ is provable iff r represents the formula obtained by writing a left parenthesis followed by the formula represented by s followed by the material conditional sign followed by the formula represented by t followed by the right parenthesis.

(7) There is a two place function expression Q such that $Q(n, s) = t$ is provable iff t represents the result of universally quantifying the nth variable in the formula represented by s.

(8) There is a one place predicate expression Sent such that Sent(s) is provable iff s represents a sentence of OL.

THEOREM. *N adequately represents its own syntax.*

Proof. We must show that each of the expressions is definable in N using the Gödel numbering as the map from the syntax of N to terms of N. Subst was already defined in Chapter VII. Var can be defined as $x = 47 \cdot 53^n$. Con(x, n) is defined as $x = 29 \wedge n = 1$. F1Term is defined as $i = 1 \wedge s = t_1 * 2^{31}$. Neg is defined as $t = 2^{13} * s$.

EXERCISE 1. Define $F2$Term, Pred2, and Sent. It should be clear to the reader that if we added to N any finite number of constants, function symbols and/or predicate letters we could revise the definitions given above so as to represent the syntax of the enriched language. We will assume without going through the details of the proof that

THEOREM. *N adequately represents the syntax of any language consisting of the result of adding a finite list of constants, function symbols and/or predicate letters to N.*

We are now in a position to prove a general version of the theorem proved for N in the earlier chapter.

THEOREM. *If* OL *is adequate for its own syntax, then there is no definition of truth in* OL *which satisfies Tarski's conditions.*

Proof. We follow the proof given earlier in a more general form. Suppose there were a theory of truth which met Tarski's conditions, i.e., there is a predicate expression T with one free variable such that $T(\text{'}A\text{'}) \leftrightarrow A$ is provable for every A in the language. We let v be the free variable of T and form a new expression by substituting the function expression Subst$(v, \text{'}v\text{'}, v)$ for v in T and negating the resulting expression, i.e., $-T(\text{Subst}(v, \text{'}v\text{'}, v))$. This formula is represented by a term which we will call t. Consider the formula $-T(\text{Subst}(\text{'}t\text{'}, \text{'}v\text{'}, \text{'}t\text{'}))$. It is represented by some term s and since the formula is the result of substituting the term representing 't' for v in the formula t, $\vdash s = \text{Subst}(\text{'}t\text{'}, \text{'}v\text{'}, \text{'}t\text{'})$. Since the theory meets Tarski's condition $\vdash T(s) \leftrightarrow -T(\text{Subst}(\text{'}t\text{'}, \text{'}v\text{'}, \text{'}t\text{'})$, and so $\vdash T(s) \leftrightarrow = T(s)$, and the theory is inconsistent.

COROLLARY. *No finite extension of N is such that a theory meeting Tarski's conditions can be given for the language in the same language.*

Proof. By the previous two theorems.

EXERCISE 2. Prove the theorem by using the result in Exercise 10 of Chapter VIII.

If we are to prove the theorems mentioned in the Tarski condition we must also assume that ML contains the non-logical expressions of OL, i.e., that each constant, function symbol and predicate letter of OL is also in ML. We will show that with this assumption and the assumption that ML represents the syntax of OL we can define truth in ML by adding three new expressions and a list of axioms to be given. In giving this definition we will be mirroring the definition of truth in a model given in Chapter I, with the slight difference that we will use the atomic expressions of ML to specify the interpretation. In particular, we will 'interpret' a predicate F of OL as being true of all and only the objects in the domain of OL which satisfy F in ML. The expressions which we add are a one place predicate D which characterizes the domain of OL, a one place predicate Seq which intuitively is to be interpreted as being true of sequences and a two-place function symbol Den(x, t) which will be interpreted as specifying the object which is the denotation of t relative to the sequence x.

Before giving the axioms which are required for the truth theory, it will be useful to introduce two definitions. We will write $x(t)$ for Den(x, t), i.e., $x(t)$ will be the denotation of t relative to the sequence x. We will also use the formal version of the abbreviation $\alpha \overset{v}{\underset{x}{\approx}} \beta$ used in earlier chapters. Thus we will use $x \overset{n}{\underset{y}{\approx}} z$ as an abbreviation of the expression $\mathrm{Seq}(x) \wedge \mathrm{Seq}(z) \wedge (w)(\mathrm{Var}(w, n) \wedge z(w) = y) \supset (u)(m)(m \neq n \supset (\mathrm{Var}(u, m) \supset x(u) = z(u)))$. Informally, this is to say that x and z are sequences which agree on every variable except possibly the nth and that z assigns y to the nth variable. We shall also use $x \overset{n}{\approx} z$ as an abbreviation of $(Ey)(x \overset{n}{\underset{y}{\approx}} z)$.

Our first few axioms guarantee that there are sequences, that sequences are total and that for any object in D and any variable and any sequence there is another sequence which is like the first except (possibly) that it assigns the object specified to the variable specified.

(T1) $(Ex)\mathrm{Seq}(x)$

(T2) $(x)(y)(n)(Ez)[\mathrm{Seq}(x) \wedge \mathrm{Var}(y, n) \supset z = x(y) \wedge Dz]$

(T3) $(x)(y)(z)(n)(Eu)[\mathrm{Seq}(x) \wedge D(y) \wedge \mathrm{Var}(z, n) \supset u \overset{n}{\underset{y}{\approx}} x]$.

Next we must add axioms to ensure that sequences assign the appropriate objects to terms other than variables. We must do this by cases, so first we add axioms for constants:

(T4n) $(x)(y)[\mathrm{Seq}(x) \wedge \mathrm{Con}(y, n) \supset x(y) = c]$ where c is the nth constant.

For each n place function symbol there will be a similar axiom:

(T5ni) $(x)(y)(t_1) \ldots (t_n)[\mathrm{Seq}(x) \wedge Fn\mathrm{Term}(i, t_1, \ldots t_n) = y \supset x(y) = f_1^n(x(t_1) \ldots x(t_n))]$.

Finally, we must have a two-place relation Sat to formalize the satisfaction relation and we add axioms to characterize it. The first axioms are those corresponding to the atomic predicates. For each n-place atomic predicate (including identity as a two place predicate) we have an axiom

(T6ni) $(x)(y)(t_1) \ldots (t_n)[\mathrm{Seq}(x) \wedge \mathrm{Pred}n(i, t_1, \ldots t_n) = y \supset \mathrm{Sat}(x, y) \leftrightarrow F_i^n(x(t_1) \ldots x(t_n))]$.

And finally we add the axioms characterizing satisfaction for complex formulas in terms of the satisfaction of their parts.

(T7) $(x)(y)(z)[\mathrm{Seq}(x) \wedge y = \mathrm{Neg}(z) \supset (\mathrm{Sat}(x, y) \leftrightarrow -\mathrm{Sat}(x, z))]$

(T8) $(x)(y)(z)(u)[\mathrm{Seq}(x) \wedge y = \mathrm{Imp}(z, u) \supset (\mathrm{Sat}(x, y) \leftrightarrow (\mathrm{Sat}(x, z) \supset \mathrm{Sat}(x, u)))]$.

(T9) $(x)(y)(z)(n)[\mathrm{Seq}(x) \wedge Q(n, z) = y \supset [\mathrm{Sat}(x, y) \leftrightarrow (u)(u \overset{n}{\approx} x \supset \mathrm{Sat}(u, z))]]$.

Finally, we introduce our truth predicate $T(y)$ as an abbreviation for $(x)[\mathrm{Seq}(x) \supset \mathrm{Sat}(x, y)] \wedge \mathrm{Sent}(y)$. Note that for a language with c constants, f function symbols and F predicates, the definition of truth will require $c + f + F + 6$ axioms. We are now ready to prove our main theorem, that the definition of truth just given satisfies Tarski's condition, assuming that it is consistent.

THEOREM. *If* OL *is a first order language and* ML *is adequate to represent the syntax of* OL *and the non-logical expressions of* OL *are included among those of* ML *and there are expressions of* ML *such that the axioms* (T1)–(T9) *are provable in* ML *and* ML *is consistent, then* ML *satisfies Tarski's condition for* OL.

Proof. We must first give a more rigorous statement of Tarski's condition. For any OL sentence A we will call A_D its relativization to D in ML if every subformula of A of the form $(v)B$ has been replaced in A_D by the subformula $(v)(Dv \supset B)$. Tarski's condition is that for any OL sentence A, if a is the term representing A in ML and A_D is the relativization of A to D in ML, then $T(a) \leftrightarrow A_D$ is provable in ML.

To prove the theorem we will prove a lemma which is in fact slightly more general. We will use 'A' as a convenient representation of the term representing the formula A.

LEMMA. *If* $v_1, \ldots v_n$ *are the free variables of* A *and* $A_D x(v_1) \ldots x(v_n)$ *is the result of substituting* $x(v_1) \ldots x(v_n)$ *for* $v_1, \ldots v_n$ *in the relativization* A_D *of* A, *then we can prove in* ML $(x)[\mathrm{Seq}(x) \supset (\mathrm{Sat}(x, `A') \leftrightarrow A_D x(v_1) \ldots x(v_n))]$.

Proof. By induction on the order k of 'A'. If the sentence in question is atomic then the lemma follows immediately from the appropriate instance of (T6ni).

If A is of the form $-B$, then we know that we can prove in ML the sentence $`A' = \mathrm{Neg}(`B')$. By the induction hypothesis, since B is of order less than k, we can assume that we can prove $(x)[\mathrm{Seq}(x) \supset (\mathrm{Sat}(x, `B') \leftrightarrow B_D x(v_1) \ldots x(v_n))]$. Since we also have as a consequence of (T7) and $\vdash `A' = \mathrm{Neg}(`B')$, $(x)[\mathrm{Seq}(x) \supset (\mathrm{Sat}(x, `A') \leftrightarrow -\mathrm{Sat}(x, `B'))]$ by quantification theory $(x)[\mathrm{Seq}(x) \supset (\mathrm{Sat}(x, `A') \leftrightarrow -B_D(x(v_1) \ldots x(v_n))]$ is provable, which is what was to be shown.

EXERCISE 3. Prove the case where A is of the form $B \supset C$. If A is of the form $(v)B$ where v is the mth variable, then since B is of order less than k, we have by our induction hypothesis that

(a) $$(x)[\mathrm{Seq}(x) \supset (\mathrm{Sat}(x, `B') \leftrightarrow B_D x(v)x(v_1) \ldots x(v_n))]$$

is provable. Since the syntax of OL is represented in ML we can also prove $`A' = Q(m, `B')$. From this and (T9) it follows that we can prove in ML

(b) $$(x)[\mathrm{Seq}(x) \supset [\mathrm{Sat}(x, `A') \leftrightarrow (u)(w)(u \underset{w}{\overset{m}{\approx}} x \wedge D(w) \supset \mathrm{Sat}(u, `B'))]].$$

Instantiating the sentence (a) and substituting equivalences in the last consequent of (b) we obtain a proof of

(c) $$(x)[\mathrm{Seq}(x) \supset (\mathrm{Sat}(x\ `A') \leftrightarrow (u)(w)(u \underset{w}{\overset{m}{\approx}} x \wedge D(w) \supset B_D u(v)u(v_1) \ldots u(v_n))].$$

Since $u \underset{w}{\overset{m}{\approx}} x$ ensures that $x(v_1) = u(v_1), \ldots x(v_n) = u(v_n)$ we can prove

(d) $$(x)[\mathrm{Seq}(x) \supset (\mathrm{Sat}(x\ `A') \leftrightarrow (u)(w)(u \underset{w}{\overset{m}{\approx}} x \supset (D(w) \supset B_D u(v)x(v_1) \ldots x(v_n)))].$$

Furthermore, since $u \underset{w}{\overset{m}{\approx}} x$ entails $w = u(v)$, we can prove

(e) $$(x)[\mathrm{Seq}(x) \supset (\mathrm{Sat}(x\ `A') \leftrightarrow (u)(w)(u \underset{w}{\overset{m}{\approx}} x \supset (D(w) \supset (D(w) \supset B_D wx(v_1) \ldots x(v_n)))].$$

Trivially,

$$(w)(D(w) \supset B_D wx(v_1) \ldots x(v_n)) \supset (u)(w)(u \underset{w}{\overset{m}{\approx}} x \supset (D(w) \supset B_D wx(v_1) \ldots x(v_n))).$$

The converse is also provable using (T2) and (T3), and thus we can prove

(f) $$(x)[\mathrm{Seq}(x) \supset (\mathrm{Sat}(x, `A') \leftrightarrow (w)(D(w) \supset B_D wx(v_1) \ldots x(v_n)))].$$

The formula beginning (w) differs only by having a different bound variable from $A_D \vee x(v_1) \ldots x(v_n)$ so by one more step of substitution of equivalence the lemma is established. Having proved the lemma we now must show that the theorem follows: Suppose A is an OL sentence, then it is provable that Sent('A') and by the lemma we can prove $(x)[\mathrm{Seq}(x) \supset (\mathrm{Sat}(x, `A') \leftrightarrow A_D)]$, since A, being a sentence, has no free variables.

To prove the Tarski conditions, we first assume A_D and derive $T(`A')$. $A_D$, $\mathrm{Seq}(x) \vdash \mathrm{Sat}(x, `A')$ by the lemma and so $\vdash A_D \supset (x)(\mathrm{Seq}(x) \supset \mathrm{Sat}(x, `A')) \wedge \mathrm{Sent}(`A')$. To prove the converse we assume $T(`A')$, i.e., $(x)(\mathrm{Seq}(x) \supset \mathrm{Sat}(x`A')) \wedge \mathrm{Sent}(`A')$. By (T1) it follows that $(Ex)(\mathrm{Seq}(x) \wedge \mathrm{Sat}(x, `A'))$ and so by the lemma we can derive A_D.

We have already shown that N is adequate for the syntax of any finite extension of the language of N. We will now investigate the extent to which we can already define the expressions D, Seq and Sat in N. Since we have already coded sequences as natural numbers in Chapter VII, we might hope to be able to prove the necessary sequence axioms in N and to take D as a predicate true of all objects, e.g., $x = x$. In other words we could treat every natural number as a sequence by letting it assign to the nth term the exponent of the nth prime in the number.

Since we are treating all natural numbers as sequences we can also let $\mathrm{Seq}(x)$ be defined as $x = x$. In defining $x(t)$ we must take some care that x is defined for all variables t. Since $(x)_i$ is defined only for i less than or equal to x we must define $x(t) = n$ as $n = (x)_t \vee (t > \mathrm{gp}(x) \wedge n = 0)$. With this definition T1 and T2 are easily provable. (T3) becomes: $(x)(y)(z)(n)(Eu)[\mathrm{Var}(z, n) \supset u \overset{n}{\underset{y}{\approx}} x]$. If we choose values for x, y, z and n, then if $y - (x)_n = k$, $\mathrm{Var}(z, n) \supset x \cdot p_n^k \overset{n}{\underset{y}{\approx}} x$ will be provable. If $y < (x)_n$ then, if k is the difference, x divided by p_n^k will be a u which provably meets the condition. Thus since we can prove $y < (x)_k \vee y = (x)_k \vee y > (x)_k$, (T3) will be provable in N.

Next we must show that the denotation relation $\mathrm{Den}(x, t)$ is definable. In fact, $\mathrm{Den}(x, t)$ will be a primitive recursive function of x and t, which is defined by cases:

(D_1) $\quad (En < t)\mathrm{Var}(t, n) \wedge \mathrm{Den}(x, t) = x(t)$.

(D_2) $\quad \mathrm{Den}(x, 29) = 0$

(D_3) $(En < t)\ t = n{*}2^{31} \wedge \mathrm{Den}(x, t) = \mathrm{Den}(x, n) + 1$

(D_4) $(Em, n < t)\ \ t = \text{‘}(m + n)\text{’} \wedge \mathrm{Den}(x, t) = \mathrm{Den}(x, n) + \mathrm{Den}(x, m)$

(D_5) $(Em, n < t)\ \ t = (\text{‘}mn\text{’}) \wedge \mathrm{Den}(x, t) = \mathrm{Den}(x, m) \cdot \mathrm{Den}(x, n)$

(D_6) $(Em, n < t)\ t = \text{‘}m^{n}\text{’} \wedge \mathrm{Den}(x, t) = \mathrm{Den}(x, m)^{\mathrm{Den}(x,n)}$.

The definition is $(D_1) \vee (D_2) \vee (D_3) \vee (D_4) \vee (D_5) \vee (D_6) \vee [\mathrm{Den}(x, t) = 0 - ((D_1) \vee (D_2) \vee (D_3) \vee (D_4) \vee (D_5) \vee (D_6))]$.

With this definition, the instances of axioms (T4n) and (T5ni) for *N* are provable. Thus we can find a truth definition for *N* by simply adding a new expression Sat and axioms corresponding to (T6ni) and (T7), (T8) and (T9).

THEOREM. *There is a theory whose vocabulary exceeds that of N only by including a two place relation not in N such that that theory satisfies Tarski's conditions for defining truth for N.*

It is relatively easy to see that this theorem will generalize as did our previous theorem to theories which are finite extensions of *N*, i.e. which exceed the vocabulary of *N* by including only a further finite list of constants, function letters and predicate letters.

THEOREM. *If* OL *is a finite extension of N, then there is a truth theory satisfying Tarski's conditions for defining truth in* OL *which can be given in a theory which results by adding a new two place predicate to* OL *(together with a suitable finite set of axioms).*

Proof. Note that in proving the axioms for the theory of truth we only used finitely many axioms of *N* thus we need only those axioms plus the axioms characterizing satisfaction which will also be a finite list if OL is a finite extension of *N*.

We will now prove a theorem due to Kleene which shows that any theory which is in a vocabulary which is a finite extension of *N* and which is effectively axiomatizable can be *finitely* axiomatized if we add a new relation to the language.

THEOREM. *Let H be a theory in a language which is a finite extension of N and let the theorems of H be recursively enumerable. H is finitely axiomatizable in a language obtained by adding a new two-place predicate to the language of H.*

Proof. By the previous theorem we can define a truth predicate T_h for the language of H such that from a finite set of axioms of N we can prove $T_h(\text{'}A\text{'}) \leftrightarrow A$ for all A in the language. Since the theorems of H are recursively enumerable we know that there is a two-place relation $B(x, y)$ such that A is a theorem iff $(Ey)B(y, \text{'}A\text{'})$ is provable in Q. Let F be the conjunction of the finite set of axioms of N used in defining T_h and let $\bigwedge Q$ be the conjunction of the axioms of Q. Then the single axiom $F \wedge \bigwedge Q \wedge (x)[(Ey)B(y, x) \supset T_h(x)]$ suffices to prove all theorems of H. Let A be a theorem of H, then $(Ey)B(y, \text{'}A\text{'})$ will be provable in Q and thus from our axiom we can prove $T_h(\text{'}A\text{'})$. Now using F we can prove $T_h(\text{'}A\text{'}) \leftrightarrow A$, so from our single axiom we can prove A.

Our final theorem in this chapter will show that we can approximate a truth definition for N in N in the sense that for any k, we can define a truth predicate for sentences containing at most k quantifiers in N itself.

THEOREM. *For any k, there is a predicate T_k such that for any sentence A of N containing at most k quantifiers* $\vdash_N T_k(\text{'}A\text{'}) \leftrightarrow A$.

Proof. A sentence containing no quantifiers can only be of one of the forms $s = t$, where s and t are closed terms, or $-B$ or $B \supset C$. We define a predicate T_0 by cases for sentences without quantifiers:

$$T_0(\text{'}s = t\text{'}) \text{ iff } \mathrm{Den}(0, s) = \mathrm{Den}(0, t)$$
$$T_0(\text{'}-B\text{'}) \text{ iff } -T_0(\text{'}B\text{'})$$
$$T_0(\text{'}B \supset C\text{'}) \text{ iff } T_0(\text{'}B\text{'}) \supset T_0(\text{'}C\text{'}).$$

If x is not of one of the above forms, or contains quantifiers $-T_0(x)$. By the recursion theorem we can find an expression of N which defines T_0 since the truth value of $T_0(x)$ is always specified in terms of the values of T_0 for smaller arguments.

If we are given a definition of T_k we define T_{k+1} by cases analogously to T_0 with the addition of a quantifier clause in terms of T_k. We use $Q_{k+1}(\text{'}A\text{'})$ as an abbreviation of the predicate 'A contains at most $k + 1$ quantifiers'.

T_{k+1}('A') iff [Q_{k+1}('A') ∧ ['A' = '$(v)B$' ∧ $(x)T_k$
(Sub($\underline{x}$, 'v', 'B'))) ∨
('A' = '$-B$' ∧ $-T_{k+1}$('B')) ∨ ('A' = '$B \supset C$' ∧ T_{k+1}('B') ⊃
T_{k+1}('C')]] ∨ (Q_k('A') ∧ T_k('A')).

This recursive definition by cases can be turned into an explicit definition of T_{k+1} in terms of T_k in N. We will now show that the definition is adequate.

THEOREM. *The predicates T_k each satisfy Tarski's condition for sentences of N with at most k quantifiers.*

Proof. We prove the theorem by establishing two lemmas.

LEMMA 1. T_k('A') $\leftrightarrow A_{T_0}$, *where A_{T_0} is the sentence obtained from A by replacing the variables $v_1, \ldots v_n$* by *$x_1, \ldots x_n$ and replacing each atomic part $s = t$ by* T_0(Subst$^{v_1, v_2, \ldots v_n}_{\underline{x}_1, \underline{x}_2, \ldots \underline{x}_n}$ '$s = t$').

The proof of the lemma is by induction on k. If $k = 0$ then it is clear that if A is a truth functional sentence T_0('A') is provably equivalent to the result of replacing each atomic part B of A by T_0('B'). We now assume that the hypothesis is true for $k = n$ and show that it holds for $k = n + 1$. If A contains $n + 1$ quantifiers and is truth functionally complex, then T_k('A') is provably equivalent to the result of replacing each truth functional component B of A with T_k('B'). By the definition of T_k we know that T_k('$(v)B$') is provably equivalent to $(x)T_{k-1}$(Sub($\underline{x}$, 'v', 'B') which by the induction hypothesis is provably equivalent to $(x)B_{T_0 \underline{x}}^{\ v}$. So the lemma follows by substitution of equivalences.

LEMMA 2. $\vdash_N A_{T_0} \leftrightarrow A$.

Since A_{T_0} is like A except for containing variables $x_1, \ldots x_n$ where A contains $v_1, \ldots v_n$ and for having occurrences of T_0(Sub($^{v_1 \ldots v_n}_{\underline{x}_1 \ldots \underline{x}_n}$ '$s = t$')) where A contains $s = t$, it will suffice to prove the lemma if we can prove that for all terms s and t

$$\vdash_N (x_1) \ldots (x_n)[T_0(\text{Sub}(^{v_1 \ldots v_n}_{\underline{x}_1 \ldots \underline{x}_n}, \text{`}s = t\text{'}) \leftrightarrow s = t^{v_1 \ldots v_n}_{x_1 \ldots x_n}].$$

By the definition of T_0,

(1) $$\vdash_N (x_1)\ldots(x_n)[T_0(\mathrm{Sub}(^{v_1 \ldots v_n}_{\underline{x}_1 \ldots \underline{x}_n}, \text{`}s = t\text{'}) \leftrightarrow (\mathrm{Den}(0, \mathrm{Sub}^{v_1, \ldots v_n}_{\underline{x}_1, \ldots \underline{x}_n}, \text{`}s\text{'}) = \mathrm{Den}(0, \mathrm{Sub}^{v_1 \ldots v_n}_{\underline{x}_1 \ldots \underline{x}_n}, \text{`}t\text{'}))].$$

We will now show that for each term of the language

$$\vdash_N (x_1)(x_2)\ldots(x_n)(\mathrm{Den}(0, \mathrm{Sub}^{v_1 \ldots v_n}_{\underline{x}_1 \ldots \underline{x}_n} \text{`}s\text{'}) = s^{v_1 \ldots v_n}_{x_1 \ldots x_n})$$

from which the lemma will follow by substitution of identicals and equivalence. We note first that $\vdash \mathrm{Den}(0, \text{`}0\text{'}) = 0$ and that by $(T5_{1,1}) \vdash (x)$ $\mathrm{Den}(0, \text{`}\underline{x}'\text{'}) = \mathrm{Den}(0, \text{`}\underline{x}\text{'})'$ and thus $\mathrm{Den}(0, \text{`}\underline{x}\text{'}) = x \supset \mathrm{Den}(0, \text{`}\underline{x}'\text{'}) = x'$, and so by induction in $N$ we can prove $\vdash (x)\ \mathrm{Den}(0, \text{`}\underline{x}\text{'}) = x$. We will now prove the same theorem for terms which are not numerals by induction in the metalanguage on the complexity of the term. What we do is to show that for each degree of complexity we can prove in N the appropriate theorem. Thus we assume that for all terms of complexity less than k we can prove that if $v_1 \ldots v_n$ are all the variables occurring in t then

$$\vdash (x_1)\ldots(x_n)\mathrm{Den}(0, \mathrm{Sub}(^{v_1 \ldots v_n}_{\underline{x}_1 \ldots \underline{x}_n}, \text{`}t\text{'}) = t^{v_1 \ldots v_n}_{\underline{x}_1 \ldots \underline{x}_n}.$$

Let us now consider a term of complexity $k+1$. If the term is a successor term then we can use the same argument used above with $t^{v_1 \ldots v_n}_{\underline{x}_1 \ldots \underline{x}_n}$ replacing $\underline{x}$ in the argument. If t is an addition term, i.e., t is $r+s$, then letting r_x stand for the result of substituting $\underline{x}_1 \ldots \underline{x}_n$ for $v_1 \ldots v_n$ in r and correspondingly for s_x, we have by our induction hypothesis that $\vdash \mathrm{Den}(0, \text{`}r_{\underline{x}}\text{'}) = r_x$ and $\vdash \mathrm{Den}(0, \text{`}s_{\underline{x}}\text{'}) = s_x$. As a consequence of $(T5_{2,1})$ we have $\vdash \mathrm{Den}(0, \text{`}r_{\underline{x}} + s_{\underline{x}}\text{'}) = \mathrm{Den}(0, \text{`}r_{\underline{x}}\text{'}) + \mathrm{Den}(0, \text{`}s_{\underline{x}}\text{'})$ and so $\vdash \mathrm{Den}(0, \text{`}r_{\underline{x}} + s_{\underline{x}}\text{'}) = r_x + s_x$, which is what was to be shown.

EXERCISE 4. Prove the cases where the term is multiplicative or exponential. With the results of this exercise we have completed the proof of the claim that the closure of $\mathrm{Den}(0, \text{`}t_{\underline{x}}\text{'}) = t_x$ is provable for all terms t. Lemma 2 follows from this fact and the formula (1) above. Putting together Lemmas 1 and 2, we see that the truth definitions T_k meet Tarski's conditions for defining truth for the relevant subsystems of N.

CHAPTER X

SOME RECURSIVE FUNCTION THEORY

We defined a function f to be computable just in case the relation $f(x_1, \ldots x_n) = z$ is n.r. in Q. This means that for some number e

$$f(n_1, \ldots n_k) = m \text{ iff } \vdash \underline{(\exists\, a)T(e, n_1, \ldots n_k, m, a)}$$
$$f(n_1, \ldots n_k) \neq m \text{ iff } \vdash \underline{-T(e, n_1, \ldots n_k, m, j)} \text{ for all } j.$$

It will be useful to introduce the notation $\{e\}(n_1, \ldots n_k)$ for $f(n_1, \ldots n_k)$ where e is the number assigned to the function f. Recall that we defined a set S to be *weakly n.r.* just in case there is an Aa such that $n \in S$ iff $\vdash_Q \underline{An}$. We will now show the equivalence of several concepts of semi-effectiveness. Df S is *recursively enumerable* (r.e.) iff there is a total computable f such that $n \in S$ iff $(\exists\, z)f(z) = n$.

THEOREM 1. *S is r.e. iff it is weakly n.r.*

Proof. Suppose S is r.e. and let e be the number assigned f, then $n \in S$ iff $(\exists z)f(z) = n$ iff $\vdash (\exists a)(\exists b)\underline{T(e, b, n, a)}$; this last formula weakly n.r. represents S. Suppose S is weakly n.r. and let Aa be the formula which represents it, i.e. $n \in S$ iff $\vdash_Q \underline{An}$. Let k be the least integer in S. Then the function, $f(a) = n$ if $T(e, n, a)$ and $f(a) = k$ otherwise, will generate S if e is the number of $\underline{Aa}$. This function is total and is computable since the expression $(T(e, n, a) \wedge b = n) \vee (-T(e, n, a) \wedge b = k)$ will n.r. $f(a) = b$.

EXERCISE 1. Show that S is r.e. iff it is the domain of a partial computable function, i.e. $n \in S$ iff $(\exists\, z)f(n) = z$ for some partial computable f.

EXERCISE 2. Show that any finite set is recursively enumerable.

EXERCISE 3. Show that if S is r.e. by f and f is monotonic, then S is effective [f is monotonic if $f(x+1) > f(x)$].

EXERCISE 4. Show that every infinite r.e. set has an infinite decidable subset.

Our Gödel numbering of formulas of Q gives us an enumeration of the partial computable functions. We will now show that it is not possible to find even a recursive enumeration of the total functions. In other words, the set of complete computable functions is not r.e.

For each n there is a function $f_n(a)$ such that $f_n(a) = 0$ if $-T(n, n, a)$ and f_n is undefined otherwise. Obviously f_n is a total function iff $(b)-T(n, n, b)$. f_n is computable because $f_n(a) = b$ is n.r. by $\underline{-T(n, n, a) \wedge b = 0}$. Thus if the set of total functions were r.e. the set $\{n: (b) - T(n, n, b)\}$ would be r.e., and thus it would be weakly n.r. But we know $\{n: -(b) - T(n, n, b)\}$ is weakly n.r. and we have proved that if a set and its complement are both weakly n.r. then they are n.r. Therefore, if the set of total computable functions is r.e., the predicate $(\exists b)T(n, n, b)$ is decidable. This proves our

THEOREM 2. *The set of total computable functions is not r.e.*

This theorem leads us to consider the classification of predicates which are not recursive or recursively enumerable. The set of total computable one place functions can be defined using the T predicate as $\{n: (x)(Ey)T(n, x, y)\}$. Thus we have at least the following classes of sets:

sets definable by recursive predicates

sets definable by existential quantification of a recursive predicate, i.e., r.e.

sets definable by universal quantification of a recursive predicate, i.e., complements of r.e. sets

sets definable with two quantifiers applied to a recursive predicate.

This leads us to the following definition of a hierarchy of sets. A set is Σ_n if it can be defined by the application of n alternating quantifiers beginning with an existential quantifier applied to a recursive predicate; a set is Π_n if it can be defined by the application of n alternating quantifiers to a recursive predicate where the alternation begins with a universal quantifier; a set is Δ_n if it is both Σ_n and Π_n. Thus the recursively enumerable sets are the Σ_1 sets and their complements are the Π_1 sets and the set of computable functions is a Π_2 set.

EXERCISE 5. What sets are the Δ_1 sets?

We now want to prove that the classes of sets which we have defined are in fact distinct. The hierarchy was first defined by Kleene and thus is sometimes called the Kleene hierarchy, though it is more frequently called the arithmetic hierarchy. This term derives from the fact that any set definable in the language of number theory, i.e., N, can be placed somewhere in this hierarchy.

THEOREM 3. *Let S be a set of natural numbers definable in the vocabulary corresponding to that of N. S is a member of* Σ_n *or* Π_n *or* Δ_n *for some n.*

Proof. Let A be the expression which defines S and let A' be the prenex equivalent of A. Then A' is the result of applying some string of quantifiers to a quantifier free formula. The quantifier free formula is a recursive predicate since it is composed only out of truth functions and the functions of N and every result of substituting numerals for the variables will be provable or refutable in Q. We will now show that if A' contains any adjacent quantifiers of the same type we can eliminate all but one of these quantifiers. That is, suppose that the formula A' includes a part of the form $(x_1)(x_2)B(x_1, x_2)$. This formula is provably equivalent to $(z)B((z)_1, (z)_2)$ if z does not occur in B, and thus there is a formula equivalent to A' which does not contain that pair of adjacent quantifiers. The same argument can be applied to strings of similar quantifiers of any length and applies equally well to existential as universal quantifiers. Thus by this process of eliminating adjacent quantifiers of the same type we find an expression of either type Σ_n or Π_n.

It should be mentioned that if a predicate can be written in the form Σ_n or Π_n then it can also be written in Σ_m and Π_m and Δ_m form for all $m > n$ by the addition of inessential quantifiers. For example $(x)(Ey)Rxyz$ can be rewritten as $(Ew)(x)(Ey)(Rxyz \wedge w = w)$ or as $(x)(Ey)(w)(Rxyz \wedge w = w)$. Our next theorem will show that the converse is not true, i.e., that some predicates which are Σ_n or Π_n or Δ_n cannot be expressed as predicates of any lower type.

KLEENE'S HIERARCHY THEOREM. *There are predicates which are $\Sigma_n(\Pi_n)$ but not $\Pi_n(\Sigma_n)$. There are predicates which are Δ_n but neither Π_{n-1} nor Σ_{n-1}.*

Proof. We will show first that there are Σ_n predicates which are not Π_n. A predicate can be expressed in Σ_n form iff there is a recursive R such that the predicate is defined as $(Ex_1)(x_2)\ldots R(m, x_1, x_2 \ldots)$. Suppose in the prefix x_n is existentially quantified. Then we know that $(Ex_n)R(m, x_1, \ldots x_n)$ is true iff for some e, $(Ex_n)(Ez)T(E, m, x_1, \ldots, z)$. Thus there is a Σ_n two-place predicate which is true of a pair e, m iff m belongs to the eth Σ_n set, i.e.,

$$(Ex_1)(x_2)\ldots(x_{n-1})(Ez)T(e, m, x_1, \ldots x_{n-1}, (z)_1, (z)_2).$$

Let us call this predicate $P(e, m)$, then $P(m, m)$ is a one place Σ_n predicate and its complement $-P(m, m)$ is a one-place Π_n predicate. If $-P(m, m)$ were also a Σ_n predicate then there would be some e for which $(m)(P(e, m) \leftrightarrow -P(m, m))$, which is impossible. Therefore $-P(m, m)$ is not Σ_n and so $P(m, m)$ is not Π_n.

If n is even then in the prefix of the Σ_n predicate x_n is universally quantified, i.e., the predicate is $(Ex_1)\ldots(x_n)R(m, x_1, \ldots x_n)$. Since R is recursive so is $-R$ and if we let e characterize $-R$ then $R(m, x_1, \ldots x_n)$ iff $-(Ez)T(e, m, x_1, \ldots x_n, z)$. Thus there will again be a two place Σ_n predicate, this time

$$(Ex_1)(x_2)\ldots(Ex_{n-1})(z) - T(e, m, x_1, \ldots x_{n-1}, (z)_1, (z)_2)$$

which enumerates the one-place Σ_n predicates. The remainder of the argument is as above and again shows that $P(m, m)$ is a Σ_n predicate which is not Π_n.

In order to show that there are Δ_n sets which are not Σ_{n-1} or Π_{n-1}, we begin with a set S which is Σ_{n-1} but not Π_{n-1} and a set P which is Π_{n-1} but not Σ_{n-1}. We will show that $\{m: m = 2^z 3^w$ and $S(z)$ and $P(w)\}$ is Δ_n. This set is clearly not Σ_{n-1} since if there were a Σ_{n-1} definition of it, $B(m)$, then $(Ez)B(2^z 3^w)$ would give a Σ_{n-1} definition of $P(w)$ which is impossible. If $B(m)$ had a Π_{n-1} definition, then if k is an element of P, we can show that $S(z) \leftrightarrow (x)\ [x = 2^z \cdot 3^k \supset B(x)]$, and the expression on the right could be shown to be a Π_{n-1} predicate. Since $S(n)$ is not Π_{n-1}, we can conclude that $B(n)$ is not Π_{n-1}.

To show that $B(n)$ is Δ_n, we assume that we are given a definition of S using $x_1, \ldots x_{n-1}$ and one of P using $y_1, \ldots y_{n-1}$. Note first that $B(m) \leftrightarrow (Ew)(Ez)\ [m = 2^w \cdot 3^z \wedge S(w) \wedge P(z)]$. We can put this into prenex form as $(Ew)(Ez)(Ex_1)(y_1)(x_2)(Ey_2)(Ex_3) \ldots [m = 2^w \cdot 3^z \wedge R_s(w) \wedge R_s(w) \wedge R_p(w)]$ where R_s and R_p are the recursive predicates from which S and P are defined. This prenexed expression can be rewritten in an equivalent Σ_n form by collapsing adjacent quantifiers. On the other hand, we also could prenex the definition as

$$(y_1)(Ew)(Ez)(Ex_1)(Ey_2)(x_2)(y_3) \ldots [m = 2^w \cdot 3^z \wedge R_s(w) \wedge R_p(w)]$$

which can be rewritten in Π_n form. Thus $B(m)$ is Δ_n since it can be written in both forms.

EXERCISE 6. Give a specific example of a set which is Δ_2 but not Σ_1 or Π_1 and prove that it has those properties. Our generalized form of Gödel's theorem amounted to showing that any first order theory with a recursive set of axioms was incomplete if correct, i.e., ω-consistent. We will now generalize that theorem and show that if the set of axioms of a first order theory is definable in the arithmetic hierarchy and the theory is correct, then it is incomplete.

MOSTOWSKI'S GENERALIZATION OF GÖDEL'S THEOREM. *If T is a first order theory whose axioms are Σ_n or Π_n and whose axioms are true in the standard model of number theory, then T is incomplete.*

Proof. Let $A(x)$ be the predicate which characterizes the axioms of the theory. Then we can define a predicate analogous to Bew which will be Σ_n if $A(x)$ is Σ_n and Σ_{n+1} if $A(x)$ is Π_n. A formula is provable iff there is a sequence which is the number of a proof of it, i.e., we can define the predicate "'B' is provable in T" as

$$(Ez)(i \leqslant \mathrm{lh}(z))[A((z)_i \vee (Ej, k < i)\mathrm{ImCon}((z)_j, (z)_k, (z)_i]$$

'B' $= (z)_{\mathrm{lh}(z)}$. Since the bounded quantifiers are eliminable by replacing them with recursive predicates, the predicate defining provability can be rewritten in a form which contains one existential quantifier preceding the quantifiers of $A(x)$.

Suppose now that we have a complete and correct formal system with a set of Σ_n or Π_n axioms, then we would have a complete correct formal system with a Σ_n or Σ_{n+1} set of theorems. Let $P(x)$ be a set which is Π_{n+1} but not Σ_{n+1}. P is definable as $(x_1)(Ex_2)\ldots R(x, x_1, \ldots x_{n+1})$, with recursive R. Since R is numeralwise representable, in a formal system whose theorems are true in the standard model, if $\underline{R}$ is the predicate which numeralwise represents R, $\vdash (v_1)(Ev_2)\ldots \underline{R}(\underline{n}, v_1, v_2\ldots)$ implies $(x_1)(Ex_2)\ldots R(n, x_1, x_2\ldots)$. And if the theory is complete then we know that $\vdash (v_1)(Ev_2)\ldots \underline{R}(\underline{n}, v_1, v_2\ldots)$ iff $(x_1)(Ex_2)\ldots R(n, x_1, x_2, \ldots)$. Thus if we had a complete correct formal system we could define $P(x)$ in terms of provability in the formal system. That is, we would know that $P(n)$ iff $\vdash(v_1)(Ev_2)\ldots R(\underline{n}, v_1, v_2\ldots)$, which is characterizable by a Σ_{n+1} predicate. Since this is impossible we can infer that the formal system cannot be both complete and correct.

This theorem has shown that a Π_n set of axioms generates a set of first order theorems which are Σ_{n+1}; we will now prove the converse.

GENERALIZED CRAIG THEOREM. *If S is a Σ_{n+1} set of sentences closed under consequence, then there is a Π_n set of axioms such that the consequences of those axioms are exactly the sentences in S.*

Proof. We define '$\bigwedge_n A$' to be the sentence which consists of n conjunctions of A. Clearly 'A' is a first order consequence of '$\bigwedge_n A$'. We let the axioms of the theory be the set of sentences '$\bigwedge_n A$' such that n generates 'A'. Formally, S is a Σ_{n+1} predicate, i.e., can be written as $(Ex_1)(x_2)\ldots R(m, x_1, x_2, \ldots x_{n+1})$ with recursive R. We define the set of axioms to be the sentences '$\bigwedge_n A$' such that $(x_2) \ldots R(n, `A', x_2, \ldots x_{n+1})$. Thus we have a Π_n predicate which characterizes our axioms and it is clear that the axioms generate exactly the sentences of S as theorems.

COROLLARY. *Craig's Theorem. If S is a recursively enumerable set of sentences which is closed under consequence then there is a recursive set of axioms which generate exactly the theorems of S.*

EXERCISE 7. Let S be a Σ_{n+1} set of sentences. Show that there is a Π_n subset of S such that the consequences of S are exactly those of the subset.

EXERCISE 8. Let S be a recursively enumerable set of sentences which is closed under logical consequence and let V be a subset of the vocabulary of S. Show that there is a recursive set of axioms which generate exactly those sentences of S which contain only the vocabulary V.

Next we will prove two theorems about the formation of new computable functions from previously given ones. We defined the notation $\{e\}(x_1, \ldots x_n)$ above. We will now show how to extend the notation to allow arbitrary partial computable terms in the braces $\{\ \}$.

RECURSION THEOREM. *There is a computable f such that for any e,*

$$(x_1) \ldots (x_m)(y_1), \ldots (y_n)\{f(e)\}(x_1, \ldots x_m, y_1, \ldots y_n) = \{\{e\}(x_1, \ldots x_m)\}(x_1, \ldots y_n).$$

Proof. We know that $\{\{e\}(x_1, \ldots x_m)\}(y_1, \ldots y_n) = z$ iff $(\exists d)T \times (\mu w(T(e, x_1, \ldots x_m, w, (d)_0), y_1, \ldots y_n, z, (d)_1))$. $\{e\}(x_1, \ldots x_m)$ gives the Gödel number of a formula with free variables $a_{m+1}, \ldots a_{m+n}$. We want $f(e)$ to be a formula with free variables $a_1, \ldots a_{m+n}$. We can define $\mathrm{Sub}^{a_1 \ldots a_m}_{x_1 \ldots x_m} e$ to be the formula $\mathrm{Sub}(29 \cdot 31^{x_1}, `a_1', \mathrm{Sub}(29 \cdot 31^{x_2}, `a_2'$ $\mathrm{Sub}(\ldots \mathrm{Sub}(29 \cdot 31^{x_m}, `a_m', e) \ldots)$. Thus $f(e) = \mu w(w, b) \times T(\mathrm{Sub}^{a_1 \ldots a_m}_{x_1 \ldots x_m} e)w, b)$ is the number of a formula given values of $x_1, \ldots x_m$

and $y_1, \ldots y_n$; and by construction $\vdash (\exists c) T(f(e), x_1, \ldots x_m, y_1, \ldots y_m, z, c)$ iff $\vdash (\exists d) T(\mu w) T(e, x_1, \ldots x_m, w, (d)_0), y_1, \ldots y_n, z, (d)_1))$.

The recursion theorem allows us to put an m-place function in the braces { } applied to n arguments to form a function of $m+n$ arguments. Our next theorem allows us to fix the first m argument of an $m+n$ place function and obtain an n-place function.

Smn THEOREM. *For every m and n, there is a computable function Smn such that for all e, it is the case that*

$$\{e\}(x_1, \ldots x_m, y_1, \ldots y_n) = \{Smn(e, x_1, \ldots x_m)\}(y_1, \ldots y_n).$$

Proof. We know that $\{e\}(x_1, \ldots x_m, y_1, \ldots y_n) = z$ iff $(\exists c) T(e, x_1, \ldots x_m, y_1, \ldots y_n, z, c)$. We simply let $Smn(e, x_1, \ldots x_m) = \mathrm{Sub}^{a_1 \ldots a_m}_{x_1 \ldots x_m}\, e$ and it is not difficult to show that

$$\vdash (\exists c) T(Smn(e, x_1, \ldots x_m), y_1, \ldots y_n, z, c) \quad \text{iff}$$
$$\vdash (\exists c) T(e, x_1, \ldots x_m, y_1, \ldots y_m, z, c).$$

These two theorems have a number of important uses in recursion theory, but our applications will all concern the interpretation of intuitionistic mathematics.

CHAPTER XI

INTUITIONISTIC LOGIC

Intuitionistic logic is intended to formalize mathematical reasoning, but intuitionistic mathematical reasoning rather than classical. For example, appeal to the principle of excluded middle is not permitted unless we can *decide* which disjunct is true. An existential statement is proved only if we can *construct* an instance. Thus only the computable functions of natural numbers are permissible and only the continuous functions of reals. To see a sampling of what intuitionistic mathematics looks like I recommend you look at Heyting's *Intuitionism.*

The following explanations of the logical symbols is taken more or less verbatim from Heyting (pp. 98–103) and is given in terms of assertion and construction. Very roughly, a construction is an abstract or mental object which shows that a statement is assertable. Thus a construction can be thought of as a proof if 'proof' is taken in a suitably vague sense.

We begin with the easiest connectives:

> $A \wedge B$ can be asserted iff both A and B can be asserted
>
> $A \vee B$ can be asserted iff either A or B can be asserted
>
> $A \supset B$ can be asserted iff we possess a construction r which, joined to any construction proving A would give a construction proving B
>
> $\neg A$ can be asserted iff we possess a construction which from the supposition that a construction proving A can be carried out leads to a contradiction, in other words, a construction which proves that no proof of A is possible
>
> $(x)Ax$ can be asserted with respect to a domain D iff we possess a construction r such that given any $d \in D$ we can obtain from r a proof of $A\underline{d}$

> $(\exists x)Ax$ can be asserted w.r.t. domain D iff we possess a construction r such that r gives us a $d \in D$ and a proof that $A\underline{d}$.

Heyting has also given a set of axioms for intuitionistic logic which can be shown to be equivalent to the system obtained by dropping the axiom schema DN from our natural deduction system NDSC and adding suitable quantifier rules. For our purposes this formulation is preferable to the one given by Heyting. Thus our system of intuitionistic predicate calculus IPC will have one axiom schema and sixteen rules of inference.

$$\text{Ref } \Gamma, A \vdash A$$

$$\vdash\supset \ \frac{\Gamma, A \vdash B}{\Gamma \vdash A \supset B} \qquad \supset\vdash \ \frac{\Gamma, B \vdash C \quad \Gamma \vdash A}{\Gamma, A \supset B \vdash C}$$

$$\vdash\wedge \ \frac{\Gamma \vdash B \quad \Gamma \vdash C}{\Gamma \vdash B \wedge C} \qquad \wedge\vdash \ \frac{\Gamma, A, B \vdash C}{\Gamma, A \wedge B \vdash C}$$

$$\vdash\vee L \ \frac{\Gamma \vdash B}{\Gamma \vdash B \vee C} \quad \vdash\vee R \ \frac{\Gamma \vdash B}{\Gamma \vdash C \vee B} \quad \vee\vdash \ \frac{\Gamma, A \vdash C \quad \Gamma, B \vdash C}{\Gamma, A \vee B \vdash C}$$

$$\vdash\neg \ \frac{\Gamma, B \vdash A \quad \Gamma, B \vdash \neg A}{\Gamma \vdash \neg B} \qquad \neg\vdash \ \frac{\Gamma \vdash A}{\Gamma, \neg A \vdash B}$$

$$\vdash\forall \ \frac{\Gamma \vdash A}{\Gamma \vdash (v)A} \ \text{if } v \text{ is not free in } \Gamma \qquad \forall\vdash \ \frac{\Gamma, A_t^v \vdash B}{\Gamma, (v)A \vdash B}$$

$$\vdash\exists \ \frac{\Gamma \vdash A_t^v}{\Gamma \vdash (\exists v)A} \qquad \exists\vdash \ \frac{\Gamma, A \vdash B}{\Gamma, \exists vA \vdash B} \ \text{if } v \text{ is not free in } \Gamma \text{ or } B$$

$$\text{perm } \frac{\Gamma, A, \Delta \vdash B}{\Gamma, \Delta, A \vdash B} \qquad \text{thin } \frac{\Gamma, A, A \vdash B}{\Gamma, A \vdash B}$$

$$\text{cut } \frac{\Gamma \vdash A \quad \Delta, A \vdash B}{\Gamma, \Delta \vdash B}$$

We will define consistency as before: Γ is an inconsistent set of formulas iff for some finite $\Delta \subseteq \Gamma$ and a formula A, both $\Delta \vdash A$ and $\Delta \vdash \neg A$. Γ is consistent if it is not inconsistent.

Note that all of the rules in IPC are classical rules and thus

THEOREM I. *If* $\Gamma \vdash_{\mathrm{IPC}} A$ *then* $\Gamma \vdash_{\mathrm{PC}} A$.

COROLLARY. *If Γ is a set of formulas which is classically consistent, then it is intuitionistically consistent.*

We will now give a partial converse of Theorem I. We will show later that $\neg\neg A \supset A$ is not provable in IPC so the best we can do is partial converses. We will define inductively a formula A^* in IPC associated with each formula A in PC. If A is atomic A^* is A. If A is $\neg B$, $B \wedge C$, $B \supset C$ or $(v)B$ then A^* is $\neg B^*$, $B^* \wedge C^*$, $B^* \supset C^*$ or $(v)B^*$. If A is $B \vee C$ then A^* is $\neg(\neg B^* \wedge \neg C^*)$; if A is $(\exists v)B$ then A^* is $\neg(v)\neg B^*$. We let $\Gamma^* = \{A^*\colon A \in \Gamma\}$.

THEOREM II. *If* $\Gamma \vdash_{\mathrm{PC}} A$ *then* $\Gamma^* \vdash_{\mathrm{IPC}} A^*$.

Proof. The idea of the proof is to show that given a derivation of $\Gamma \vdash A$ in PC we can construct a derivation of $\Gamma^* \vdash_{\mathrm{IPC}} A^*$. The proof will be by induction on the length of the derivation, but first we will prove some useful lemmas.

LEMMA 1. *For all* A, $A \vdash_{\mathrm{IPC}} \neg\neg A$.

$$\frac{\begin{array}{ll} \textit{Proof}\colon\ A \vdash A \quad \text{ref.} & A \vdash A \quad \text{ref.} \\ A, \neg A \vdash B & A, \neg A \vdash \neg B \end{array}}{A \vdash \neg\neg A.}$$

LEMMA 2. *For all* A, $\neg\neg\neg A \vdash_{\mathrm{IPC}} \neg A$.

Proof.
$$\frac{\begin{array}{ll} A \vdash \neg\neg A \ \text{Lemma 1} & A \vdash \neg\neg A \\ \neg\neg\neg A, A \vdash B \quad \neg\vdash & \neg\neg\neg A, A \vdash \neg B \end{array}}{\neg\neg\neg A \vdash \neg A.}$$

LEMMA 3. *For all* A, $A \vee \neg A \vdash_{\mathrm{IPC}} \neg\neg A \supset A$

Proof.
$$\frac{\begin{array}{ll} A, \neg\neg A \vdash A & \neg A \vdash \neg A \\ & \neg A, \neg\neg A \vdash A \quad \neg\vdash \end{array}}{\begin{array}{l} A \vee \neg A, \neg\neg A \vdash A \quad \vee\vdash \\ A \vee \neg A \vdash \neg\neg A \supset A \quad \vdash\supset \end{array}}$$

LEMMA 4. *For any A,* $\neg\neg A^* \vdash_{IPC} A^*$.

Proof. By induction on the number of connectives in A. If A is atomic, the lemma claims $\neg\neg\neg\neg A \vdash_{IPC} \neg\neg A$, which is true by Lemma 2.

For the induction step, we assume $\neg\neg B^* \vdash_I B^*$ and $\neg\neg C^* \vdash_I C^*$ and we must show:

(a)	If $A = \neg B$,	$\neg\neg\neg B^* \vdash_I \neg B^*$
(b)	If $A = B \wedge C$	$\neg\neg(B^* \wedge C^*) \vdash_I B^* \wedge C^*$
(c)	If $A = B \vee C$	$\neg\neg\neg(\neg B^* \wedge \neg C^*) \vdash_I \neg(\neg B^* \wedge \neg C^*)$
(d)	If $A = B \supset C$	$\neg\neg(B^* \supset C^*) \vdash_I B^* \supset C^*$
(e)	If $A = (v)B$	$\neg\neg(v)B^* \vdash_I (v)B^*$
(f)	If $A = \exists vB$	$\neg\neg\neg(v)\neg B^* \vdash_I \neg(v)\neg B^*$

(a), (c) and (f) are immediate by Lemma 3.

EXERCISE 1. Do cases (b), (d) and (e). (*Hint*: it is useful to show that if Γ, $A \vdash_I B$ then Γ, $\neg\neg A \vdash_I \neg\neg B$, and $\Gamma, A \vdash_I B$ then Γ, $\neg B \vdash_I \neg A$).

Main proof. We will now show by induction that if $\Gamma \vdash_{PC} A$ then $\Gamma^* \vdash_{IPC} A^*$. If the derivation is of length 1, then it is either Γ, $A \vdash A$ or $\Gamma, \neg\neg A \vdash A$; the first is an axiom of IPC and Lemma 4 takes care of the other case. For the induction step we will assume the hypothesis for shorter proofs and show that the last step in the classical proof can be justified in the intuitionistic system for * formulas. The only rules that differ are $\vdash\vee$, $\vee\vdash$, $\vdash\exists$ and $\exists\vdash$ so we need consider only those cases.

Case $\vdash\vee$. If Γ^*, B^* then $\Gamma^* \vdash \neg(\neg B^* \wedge \neg C^*)$.

Proof. $\Gamma^*, B^*, \neg B^*, \neg C^* \vdash B^*$ $\Gamma^*, B^*, \neg B^*, \neg C^* \vdash \neg B^*$

$\Gamma^*, B^*, (\neg B^* \wedge \neg C^*) \vdash B^*$ $\Gamma^*, B^*, (\neg B^* \wedge \neg C^*) \vdash \neg B^*$ $\Gamma^*, B^* \vdash \neg(\neg B^* \wedge \neg C^*)$ so by cut

if $\Gamma^* \vdash B^*$ then $\Gamma^* \vdash \neg(\neg B^* \wedge \neg C^*)$.

Case $\vee\vdash$. If Γ^*, $B^* \vdash D^*$ and Γ^*, $C^* \vdash D^*$ then

$$\Gamma^*, \neg(\neg B^* \wedge \neg C^*) \vdash D^*.$$

If $\Gamma^*, B^* \vdash D^*$ then $\Gamma^*, \neg D^* \vdash \neg B^*$ and if $\Gamma^*, C^* \vdash D^*$ then $\Gamma^*, \neg D^* \vdash \neg C^*$, so $\Gamma^*, \neg D^* \vdash \neg B^* \wedge \neg C^*$ so we know that $\Gamma^*, \neg(\neg B^* \wedge \neg C^*) \vdash \neg\neg D^*$ and our result follows by Lemma 4 and cut.

Case $\vdash\exists$. If $\Gamma^* \vdash B^*$, then $\Gamma^* \vdash (\exists vB)^*$ i.e. $\Gamma^* \vdash \neg(v)\neg B^*$.

$$\frac{\begin{array}{ll}\Gamma^*, \neg B^* \vdash \neg B^* & \\ \Gamma^*, (v)\neg B^* \vdash \neg B^* & \Gamma^*, (v)\neg B^* \vdash B^*\end{array}}{\Gamma^* \vdash \neg(v)\neg B^*} \quad \begin{array}{l}\text{thinning of}\\ \text{hypothesis}\end{array}$$

Case $\exists\vdash$. If $\Gamma^*, A^* \vdash B^*$ and v is not free in Γ or B,

$$\Gamma^*, \neg(v)\neg A^* \vdash B^*.$$

EXERCISE 2. Prove this.

RELATIVE CONSISTENCY THEOREM. *N is consistent iff IN (intuitionistic N) is consistent.*

Proof. If $\vdash_{IN} A$ and $\vdash_{IN} \neg A$ then by Theorem I (p. 107) $\vdash_N A$ and $\vdash_N - A$. If $\vdash_N A$ and $\vdash_N - A$ then for some $\Gamma \subseteq N$, $\Gamma \vdash_{PC} A$ and $\Gamma \vdash_{PC} - A$ so by Theorem II, we know that $\Gamma^* \vdash_{IPC} A^*$ and $\Gamma^* \vdash_{IPC} (\neg A)^*$ but since $(\neg A)^*$ is $\neg A^*$, $\Gamma^* \vdash_{IPC} A^*$ and $\Gamma^* \vdash_{IPC} \neg A^*$. We will show now that for each axiom B of IN, $\vdash_{IN} B^*$, which will show that if $\vdash_N A$ and $\vdash_N - A$ then $\vdash_{IN} A^*$ and $\vdash_{IN} \neg A^*$. The only axiom which changes form under the * transformation is (10) which becomes $(a)\neg(b)\neg[a \neq 0 \supset a = b']$. We will show that in general $(x)(\exists y)B \vdash_{IN} (x)\neg(y)\neg B$.

$$\frac{\begin{array}{ll}B, \neg B \vdash \neg B & \\ B, (v)\neg B \vdash \neg B & B, (v)\neg B \vdash B\end{array}}{B \vdash \neg(v)\neg B}$$

$$\begin{array}{ll}\exists vB \vdash \neg(v)\neg B & \exists\vdash \\ (x)\exists vB \vdash \neg(v)\neg B & \forall\vdash \\ (x)(\exists v)B \vdash (x)\neg(v)\neg B & \vdash\forall.\end{array}$$

You should note that our proof of Theorem II was constructive in the sense that we could give an explicit effective method for getting from the derivation $\Gamma \vdash_{PC} A$ to the derivation of $\Gamma^* \vdash_{IPC} A^*$. This means that our proof is intuitionistically acceptable and so any intuitionistic criticism of classical principles must be on grounds other than consistency.

Thus far we have been proving similarities between PC and IPC, now we shall begin to show that there are differences. (Remember that just because we dropped the axiom $\neg\neg A \vdash A$ does not show that it is not derivable.) We will give a particular interpretation of the intuitionistic formulas called 'recursive realizability' and show that all formulas provable in *IN* are recursively realizable and that some instances of $A \vee \neg A$, $\neg\neg A \vee A$ and so on are *not* realizable, thus showing that those formulas are not derivable. This interpretation is fairly faithful to the intended interpretation, though it does not correspond exactly.

We will use computable (recursive) functions as a model for constructions and by using our enumeration of such functions we will be able to make sense of applying functions to functions. (Recall this is required in the interpretation of $\supset$.) It can be shown that for formulas without quantifiers or variables, i.e. containing only $=, 0, '$, $+, \times$, and exp we can effectively find a proof or disproof in N and *IN*. Thus for such formulas we can let the 'construction' be the proof in N if it exists. For formulas with free variables the realization function will depend on the values assigned to the free variables. In the definition below we will let $x_1, \ldots x_n$ be the free variables in B and $y_1, \ldots y_m$ be the free variables in C and will abbreviate $x_1, \ldots x_n$ as x and $y_1 \ldots y_m$ as y, and x, y are the free variables of A.

DEFINITION. e recursively realizes (r.r.) A iff

(1) A contains no variables and $\text{Bew}_N(e, \text{'}A\text{'})$

(2) For every assignment of numbers to x and y, $\{e\}(x, y)$ r.r. A

(3) A is $\neg B$ and for any n which r.r. $B\{e\}(n, x)$ r.r. $0 \neq 0$

(4) A is $B \wedge C$ and $\{(e)_1\}(x)$ r.r. B and $\{(e)_2\}(y)$ r.r. C

(5) A is $B \vee C$ and $\{(e_0)_1\}(x, y) = 0$ and

$\{(e)_2\}(x)$ r.r. B or $\{(e)_1\}(x, y) = 1$ and $\{(e)_2\}(y)$ r.r. C

(6) A is $B \supset C$ and for any n such that n r.r. B, $\{e\}(n, x, y)$ r.r. C

(7) A is $\forall x_i B$ and for all k, $\{e\}(x_1, \ldots x_{i-1}, k, x_{i+1}, \ldots x_n)$ r.r. B

(8) A is $\exists x_i B$ and $\{(e)_2\}(x_1, \ldots x_{i-1}, (e)_1, x_{i+1}, \ldots x_n)$ r.r. B.

We will say that if $\{B_1, \ldots B_n\} = \Gamma$ and z are the free variables in Γ, then e r.r. $\Gamma \vdash A$ iff for any $b_1, \ldots b_n$ if $b_1, \ldots b_n$ r.r. $B_1, \ldots B_n$ then $\{e\}(b_1, \ldots b_n, z)$ r.r. A. We will now prove two theorems which together show that if $\vdash_{IN} A$ then we can find an e such that e r.r. A.

THEOREM III. *If* $\Gamma \vdash_{\text{IPC}} A$ *then there is an e such that e* r.r. $\Gamma \vdash A$.

Proof. The proof will be by induction on the length of the derivation of $\Gamma \vdash_{\text{IPC}} A$. In what follows we will let $d_1, \ldots d_k$, abbreviated d, be realizations of the formulas in Γ, z will be the free variables of Γ, w those of A, x those of B and y those of C; a, b, c will be realizations of A, B, C where relevant.

$n = 1$. We must show that $\Gamma, A \vdash A$ is r.r. The function $f(d, z, a, w) = a$ suffices.

$n = k + 1$. We must show for each rule that if we can realize the premise(s) then we can find a realization of the conclusion. The rules perm and thin are trivial since we need only switch argument places. For cut let e be realizations of Δ and v the free variables of Δ. We must show that given realizations r_1 of $\Gamma \vdash A$ and r_2 of $\Delta, A \vdash B$ we can realize $\Gamma, \Delta \vdash B$. Thus $\{r_1\}(d, z)$ will realize A given realizations of Γ, so $\{r_2\}(e, v, \{r_1\}(d, z))$ will realize B given realizations of Γ and Δ. By the recursion theorem there will be an n such that $\{n\}(e, v, d, z) = \{r_1\}(e, v, \{r_1\}(d, z))$ and n will realize the conclusions

$\vdash \supset$ If e is a function of $m + 1$ arguments which realizes $\Gamma, A \vdash B$, i.e. $\{e\}(d, z, a) = b$, then $S_{m,1}(e, d, z)$ will realize $\Gamma \vdash A \supset B$.

$\vdash \wedge$ If e_1 realizes $\Gamma \vdash B$ and e_2 realizes $\Gamma \vdash C$ then $2^{e_1} \cdot 3^{e_2}$ r.r. $\Gamma \vdash B \wedge C$.

$\vdash \vee$ L If e r.r. $\Gamma \vdash B$, then $2^0 \cdot 3^e$ r.r. $\Gamma \vdash B \vee C$.

$\vdash \vee$ R If e r.r. $\Gamma \vdash B$ then $2^1 \cdot 3^e$ r.r. $\Gamma \vdash C \vee B$.

$\vdash \neg$ If e_1 r.r. $\Gamma, B \vdash A$ and e_2 r.r. $\Gamma, B \vdash \neg A$ then there are no r.r. of Γ and B together so $f(d, z) = 0$ will r.r. $\Gamma \vdash \neg B$. (Note that this assumes consistency.)

$\vdash \forall$ If e r.r. $\Gamma \vdash A$ then $\{\{e\}(d, z)\}(w)$ realizes A for any given assignment to w. To realize $\Gamma \vdash \forall vA$ we need only use the S_{mn} theorem to find a function of v which will give the function of the other variables.

$\vdash \exists$ If e r.r. $\Gamma \vdash A^v_t$ and Γ is not r.r. then $f(d, z) = 0$ will realize $\Gamma \vdash \exists vA$. If Γ is realizable then $g(w, d, z) = \mu_v\{\{e\}(d, z)\}(w, v)$ realizes A^v_t, then $f(d, z) = 2^8 \cdot 3^{\{e\}(d, z)}$ will realize $\Gamma \vdash \exists vA$.

$\supset \vdash$ Suppose r_1 r.r. $\Gamma, B \vdash C$ and r_2 r.r. $\Gamma \vdash A$ and v r.r. $A \supset B$. Then $\{v\}(\{r_2\}(d, z), w)$ will r.r. $\Gamma \vdash B$ and so $\{r_1\}(d, z, x, \{v\}(\{r_2\}(d, z, w)))$ will r.r. $\Gamma, A \supset B \vdash C$ and by the recursion theorem we can find an e s.t. $\{e\}(d, z, x, w, v)$ which is identical to the above function. If $\{e\}(d, z, a, b)$ r.r. $\Gamma, A, B \vdash C$ then $\{e\}(d, z, (x)_1, (x)_2)$ will realize $\Gamma, A \supset B \vdash C$.

$\vee \vdash$ Suppose e_1 r.r. $\Gamma, A \vdash C$ and e_2 r.r. $\Gamma, B \vdash C$. Recall that a realization of $A \vee B$ is $2^0 3^a$ or $2^1 3^b$. Thus we can define an f that r.r. $\Gamma, A \vee B \vdash C$ as

$$f(d, z, v) = \begin{matrix} \{e_1\}(d, z, (a)_2), & \text{if } (a)_1 = 0 \\ \{e_2\}(d, z, (a)_2), & \text{if } (a)_1 = 1 \end{matrix}.$$

$\neg \vdash$ If e r.r. $\Gamma \vdash A$, then either there is no realization of Γ or none of Γ and $\neg A$, so $f(d, z, a, w) = 0$ will r.r. $\Gamma, \neg A \vdash B$.

$\forall \vdash$ If e r.r. $\Gamma, A^v_t \vdash B$ then since a realization e_1 of $(v)A$ gives a realization of A for any assignment to w, $\{e\}(d, z, \{e_1\}(w)\}$ will r.r. $\Gamma, (v)A \vdash B$.

$\exists\vdash$ If e realizes $\Gamma, A \vdash B$ then $\{e\}(d, z, \{(a)_2\}((a)_1, w))$ will r.r. $\Gamma, \exists vA \vdash B$ since a realization a of $\exists vA$ gives a value $(a)_1$ for which $(a)_2$ r.r. A with the assignment of $(a)_1$ to v.

This concludes our proof by induction that if $\Gamma \vdash_{\mathrm{IPC}} A$ then $\Gamma \vdash A$ is r.r. We will now extend this to *IN*.

THEOREM IV. *If* $\vdash_{IN} A$ *then* A *is r.r.*

Proof. Since we have Theorem III we need only show that all axioms of *IN* are r.r. The proof is trivial for all the axioms except (7) and (17*N*). For example, a r.r. of 12 is a function of m and n which will give a proof $m + n' = (m + n)'$ in *IN*, but such a proof can be obtained by

$$\text{ref.}\quad m + n' = (m + n)' \vdash m + n' = (m + n)'$$
$$(b)[m + b' = (m + b)'] \vdash m + n' = (m + n)'$$
$$(a)(b)[a + b' = (a + b)'] \vdash m + n' = (m + n)'.$$

For (7) we need only note that for any m and n, if $\underline{m = n}$ is r.r. then $m = n$ and so any realization of A^a_n will realize A^a_m. For (17*N*), suppose a realizes $A0$ and that e realizes $[(a)(A \supset A^a_{a'})]$, then $\{e\}(0)$ realizes $\underline{A0 \supset A1}$ and in general $\{e\}(n)$ realizes $\underline{An \supset An'}$, and thus $\{\{e\}(0)\}(a)$ will realize $A1$ if a realizes $A0$ and so in general $\{\{e\}(n)\}(\{\{e\}(n-1)\}(\{\{e\}(n-2)\}) \ldots \{\{e\}(0)\}(a)$ will r.r. An. In other words, we let $f(e, a, n)$ be defined by $f(e, a, 0) = a$ and $f(e, a, n') = \{\{e\}(n')\}(f(e, a, n))$, which can be shown to be n.r. in the same way that the primitive recursive functions were shown to be n.r. (p. 60).

The two theorems give us our main results about realizability.

COROLLARY 1. *If* $\vdash_{IN} \exists y A x_1 \ldots x_n y$ *and* A *contains no quantifiers, then there is a computable function* f *such that for any* $x_1, \ldots x_n$ $\vdash_{IN} A x_1 \ldots x_n f(x_1, \ldots x_n)$, *or equivalently*, $f(x_1 \ldots x_n) = k$ *and* $\vdash_{IN} A x_1 \ldots x_n k$.

COROLLARY 2. $(x)[(y)\neg Txxy \vee \exists z Txxz]$ *is not provable in IN.*

Proof. This formula is not realizable because if e did realize it we would know that $(\{e\}(x))_0 = 1$ iff $(\exists z)Txxz$ and we would have a decision procedure for $\exists zTxxz$.

COROLLARY 3. *$\neg\exists yTxxy \vee \exists zTxxz$ is not provable in IN.*

Proof. We can show that $\neg\exists yA \vdash (y)\neg A$:

$$\dfrac{\begin{array}{ll} \neg(\exists y)A, A \vdash A \quad \text{ref.} & \\ \neg(\exists y)A, A \vdash \exists yA & \neg(\exists y)A, A \vdash \neg\exists yA \quad \text{ref.} \end{array}}{\begin{array}{l} \neg\exists yA \vdash \neg A \qquad \vdash\neg \\ \neg\exists yA \vdash (y)\neg A \end{array}}$$

so if the formula in question were provable the formula in Corollary 2 would be also.

COROLLARY 4. *$\neg\neg A \vdash A$ is not derivable for all A in IN.*

Proof. Corollary 3 gives a B such that $B \vee \neg B$ is *not* provable; we will show that $\vdash_{IN} \neg\neg(B \vee \neg B)$ and thus we know that $\neg\neg(B \vee \neg B) \vdash B \vee \neg B$ is *not* derivable.

$$\dfrac{\dfrac{\begin{array}{ll} \neg(B \vee \neg B), B \vdash B & \\ \neg(B \vee \neg B), B \vdash B \vee \neg B & (B \vee B), B \vdash \neg(B \vee \neg B) \end{array}}{\begin{array}{ll} \neg(B \vee \neg B) \vdash \neg B & \\ \neg(B \vee \neg B) \vdash B \vee \neg B & \neg(B \vee \neg B) \vdash \neg(B \vee \neg B) \end{array}}}{\vdash \neg\neg(B \vee \neg B)}$$

COROLLARY 5. *$\neg(x)[(y)\neg Txxy \vee \exists zTxxz]$ is realizable.*

Proof. The formula without the negation is unrealizable so 0 will realize the formula.

COROLLARY 6. *There is a formula B such that $\neg(x)B \vdash (\exists x)\neg B$ is not provable in IN.*

EXERCISE 3. Prove this corollary. (*Hint*: Use Corollary 5 and show that $\neg(B \vee C) \vdash \neg B \wedge \neg C$.)

EXERCISE 4. In the Gentzen system GPC $--A \to A$ is derived from other principles, i.e. $A \to A$, and

$$\to A, -A$$
$$--A \to A$$

excluded middle is derived from the second line of this proof. This shows that to obtain an intuitionistic Gentzen system the rule (p. 22) $\to \neg$ must be restated as

$$\Gamma, A \to$$
$$\Gamma \to \neg A.$$

The rule $\to \forall$ must also be restricted because it permits derivation of $(x)(Ax \vee B) \to (x)Ax \vee B$ if x is not free in B, but this is not intuitionistically correct. [To see this, show that $(x)(y)[-Txxy \vee \exists zTxxz]$ is realizable.] Thus the rule must be

$$\Gamma \to A$$
$$\Gamma \to (v)A,$$

i.e. in both cases we require that the Δ mentioned in the classical case is empty. Show that if in the system IGPC just described $\Gamma \to \Delta$ is derivable then $\Gamma \vdash \bigvee \Delta$, where $\bigvee \Delta$ is the disjunction of the formulas in Δ.

CHAPTER XII

SECOND ORDER LOGIC

One natural extension of the type of logic which we have been studying is to include quantification over sets of the objects in the domain. Predicate letters have been interpreted as sets (of n-tuples) of objects from the domain. When we have wanted to assert that some fact holds for all subsets of the domain, as in mathematical induction, we have used axiom schemata which require that the fact be true at least for all subsets definable in the language. Of course, for a denumerable domain there are non-denumerably many subsets whereas there are always at most denumerably many distinct definable subsets. Thus we might expect that second order logic would provide more expressive power than first order logic and this expectation is correct.

The vocabulary of second order logic will include that of first order logic without identity but will also contain an infinite list of predicate variables G_i^n, $i = 0, 1, 2, \ldots$, for every $n \geqslant 0$. Thus the predicate letters F_i^n will be predicate constants analogous to the individual constants c_i. The formation rules are the same as those for first order logic with the further clause that if A is a formula then $(\forall G_i^n)A$ is also a formula. We will also extend the terminology of bound and free occurrences of a variable to include predicate variables.

In order to give the definition of truth in a model we need to extend the use of sequences for we must now assign things to the predicate variables as well as the individual variables. We will use the notation D^n to stand for the set of all n-tuples which can be formed out of D, and we will use $\mathscr{P}(D^n)$ for the set of all sets of such n-tuples, i.e., $X \in \mathscr{P}(D^n)$ iff $X \subseteq D^n$. We do not need to extend the concept of a model since the specification of $\langle D, I\rangle$ already fixes the interpretation of all constants, both predicate and individual, and the range of the variables G_i^n will be $\mathscr{P}(D^n)$. Our sequences for second order model theory will be 'longer' than those for first order were for they must consist of an infinite sequence of sequences. For each type of variable and constant we will have a subsequence assigning entities to each of the variables or constants of that type, and we will have infinitely many such subsequences. Alternatively, we could simply drop the talk of sequences and of ordering and simply use a function which assigns an element of D to each x_i and c_i and an element of $\mathcal{P}(D^n)$ to each G_i^n and F_i^n.

We will develop the system of second order logic as an extension of HPC (Chapter II) and thus the primitive logical vocabulary will simply consist of $-$, $\supset$ and $\forall$. We will abbreviate $(\forall G_i^n)$ as (G_i^n), and we will continue to use α, β, and so on for sequences understood now in the extended sense just discussed. Also as before we will require that $\alpha(c_i) = I(c_i)$ and we will further require that $\alpha(F_i^n) = I(F_i^n)$. The definition of satisfaction will be the natural extension of the earlier definition to include the new cases where a predicate variable appears.

The relation *α satisfies A in $\langle D, I\rangle$* is defined recursively:

> α satisfies $F_i^n t_1 \ldots t_n$ in $\langle D, I\rangle$ iff $\langle \alpha(t_1), \ldots \alpha(t_n)\rangle \in I(F_i^n)$
>
> α satisfies $G_i^n t_1 \ldots t_n$ in $\langle D, I\rangle$ iff $\langle \alpha(t_1) \ldots \alpha(t_n)\rangle \in \alpha(G_i^n)$
>
> α satisfies $-A$ in $\langle D, I\rangle$ iff α does not satisfy A in $\langle D, I\rangle$
>
> α satisfies $A \supset B$ in $\langle D, I\rangle$ iff α does not satisfy A in $\langle D, I\rangle$ or α does satisfy B in $\langle D, I\rangle$
>
> α satisfies $(x_i)A$ in $\langle D, I\rangle$ iff for every β, if $\beta \underset{x_i}{\approx} \alpha$, β satisfies A in $\langle D, I\rangle$
>
> α satisfies $(G_i^n)A$ in $\langle D, I\rangle$ iff for every β, if $\beta \underset{G_i^n}{\approx} \alpha$, β satisfies A in $\langle D, I\rangle$.

A formula A is *true in D, I* iff A is satisfied in $\langle D, I\rangle$ by all sequences of $\langle D, I\rangle$. A formula is *valid* iff it is true in all models. We will again use $\models A$ for 'A is valid'. A formula A *is satisfiable* iff it is satisfied in some model $\langle D, I\rangle$ by some sequence of that model. A set of formulas is *simultaneously satisfiable* iff there is an α and a model such that α satisfies all of the formulas in that model. A *formula A is a semantic consequence* of Γ iff every sequence and model that simultaneously satisfy Γ satisfy A also. This is equivalent to saying that $\Gamma \cup \{-A\}$ is not simultaneously satisfiable. We will often express that A is a consequence of Γ by writing $\Gamma \models A$.

You may have wondered why we chose to develop second order logic as an extension of first order logic without identity rather than with identity. Our first theorem will show that it makes no difference.

IDENTITY THEOREM FOR SECOND ORDER LOGIC. *The formula* $(G_i^1)(G_i^1x \leftrightarrow G_i^1y)$ *is satisfied by* α *in* $\langle D, I\rangle$ *iff* $\alpha(x) = \alpha(y)$.

Proof. Suppose $\alpha(x) \neq \alpha(y)$, then there is some set, e.g., $\{\alpha(x)\}$ in $\mathcal{P}(D^1)$ such that if we let $\beta \underset{G_i^1}{\approx} \alpha$ and let $\beta(G_i^1)$ be $\{\alpha(x)\}$, β will not satisfy $G_i^1x \leftrightarrow G_i^1y$. If, on the other hand, $\alpha(x) = \alpha(y)$, every set which contains $\alpha(x)$ will contain $\alpha(y)$ and so the formula will be satisfied. Thus in second order logic we can define the symbol for identity: $x = y$ is an abbreviation for $(G_i^1)(G_i^1x \leftrightarrow G_i^1y)$.

We will now show that the main metatheorems of first order logic, compactness and the Löwenheim–Skolem theorem do not hold for second order logic. In order to do this it will be useful to define a special quantifier (∞x) such that $(\infty x)A$ is satisfied by α in a model just in case there are infinitely many β such that $\beta \underset{x}{\approx} \alpha$ and β satisfies A in the model. As a first step we introduce the concept of a one-to-one correlation. We will use the abbreviation $\mathrm{Corr}(G_i^2)$ for the formula $(x)(y)(z)[(G_i^2xy \supset (G_i^2xz \supset y = z)) \wedge (G_i^2yx \supset (G_i^2zx \supset y = z))]$.

LEMMA. $\mathrm{Corr}(G_i^2)$ *will be satisfied by* α *in* $\langle D, I\rangle$ *iff* $\alpha(G_i^2)$ *is a one-one correlation between two subsets of the domain.*

Proof. $\alpha(G_i^2)$ must be a set of ordered pairs and to satisfy the formula each first element must determine a unique second element and conversely.

Now we can introduce the defined quantifier (∞x) by letting the expression $(\infty x)A$ be an abbreviation for the formula $(\exists\, G_1^2)[\mathrm{Corr}(G_1^2 \wedge (\exists\, z)[A_z^x \wedge (x)(A \supset (\exists\, y)(A_y^x \wedge G_1^2xy \wedge y \neq z))]]$.

LEMMA. $(\infty x)A$ *is satisfied by* α *in* $\langle D, I\rangle$ *iff there are infinitely many* β *such that* $\beta \underset{x}{\approx} \alpha$ *and* β *satisfies* A *in* $\langle D, I\rangle$.

Proof. For the formula to be satisfied by α there must be a correlation relation which satisfies the second conjunct. Let $\alpha[A, x]$ be the set of objects d in D such that there is a $\beta \underset{x}{\approx} \alpha$ such that $\beta(x) = d$ and β satisfies A. In order for the second conjunct to be satisfied there must be an element d_1 in $\alpha[A, x]$ such that for every element d_i in $\alpha[A, x]$ there is some element of $\alpha[A, x]$ distinct from d_1 which stands in the correlation relation to d_i. Clearly this is possible if and only if $\alpha[A, x]$ is infinite, and $\alpha[A, x]$ is infinite iff the set of $\beta \underset{x}{\approx} \alpha$ such that β satisfies A is infinite.

NON-COMPACTNESS THEOREM FOR SECOND ORDER LOGIC. *There are sets of formulas Γ such that every finite subset of Γ is satisfied but Γ is not simultaneously satisfiable.*

Proof. Consider the set of sentences containing $c_i \neq c_j$ for each $i \neq j$ and also containing $-(\infty x)(x = x)$. By the previous lemma $(\infty x)(x = x)$ is satisfied by a sequence in a model iff there are infinitely many $\beta \underset{x}{\approx} \alpha$ which satisfy $x = x$. There will be infinitely many such β iff the domain is infinite since every β satisfies $x = x$. Thus $-(\infty x)(x = x)$ is satisfied by a sequence in a model iff the domain of the model is finite. Every finite subset of Γ contains at most finitely many sentences of the form $c_i \neq c_j$ and these will be simultaneously satisfied in any model which contains all of the numbers denoted by subscripts of the c_i and which assigns each c_i the number i. Thus any finite subset of Γ will be simultaneously satisfiable. But clearly all of the sentences $c_i \neq c_j$ can be satisfied only in an infinite model and $-(\infty x)(x \neq x)$ will not be satisfied in any such model.

COROLLARY 1. *There are formulas A and sets of formulas Δ of second order logic such that $\Delta \models A$, but A is not a semantic consequence of any finite subset of Δ.*

Proof. Let Δ be the set Γ of the theorem with $-(\infty x)(x = x)$ removed and let A be $(\infty x)(x = x)$.

COROLLARY 2. *The quantifier $(\infty x)A$ is not definable in first order logic.*

Proof. By the compactness theorem for first order logic and Corollary 1.

THEOREM. *There are formulas of second order logic which are satisfiable but which are not satisfiable in finite or denumerably infinite domains.*

Proof. $(\exists G_1^1)[(\infty x)G_1^1x \wedge (F_1^2) - [\mathrm{Corr}(F_1^2) \wedge (y)(\exists x)(F_1^2xy \wedge G_1^1x)]]$ is an example of such a formula. If we take the domain to be the real numbers and then we can construct a model which satisfies this formula, for the necessary value to satisfy $(\exists G_1^1)$ could be, e.g., the natural numbers. On the other hand it is clear that no finite model will satisfy the formula and if we consider a model with a denumerably infinite domain then for any infinite set we assign to G_1^1 there will be a correlation between the whole domain and the set assigned to G_1^1.

EXERCISE 1. (For readers familiar with set theory) Show that for every $\aleph_n$ there is a formula which is satisfiable in a model of cardinality $\aleph_n$ but not in a model of any smaller cardinality.

EXERCISE 2. Show that there is a formula which is satisfied in all denumerably infinite domains and in no other domains.

The falsity of the compactness theorem for second order logic implies that we cannot find a completely satisfactory set of axioms for second order logic. Any definition of derivation will require that the number of premises used in the proof be finite and thus there will be infinite sets of formulas which semantically entail contradictions but are such that we cannot deduce a contradiction from them. We will now show that the situation is even worse and that we cannot even find an effective set of axioms and rules which yield all of the semantic consequence relations between single formulas.

The valid formulas of second order logic will include all of the instances of axioms schemas of first order logic with the extended sense of formula. Furthermore, it is easy to see that modus ponens and generalization of individual variables preserve validity. The natural extensions of the quantification axiom schemata are valid and the extension of generalization to the predicate variables will preserve soundness. Thus among the valid formulas of second order logic are $(G_i^n)(A \supset B) \supset (A \supset (G_i^n)B)$, for each n, assuming G_i^n is not free in A. In order to state the universal instantiation axiom we have to define substitution for formulas and predicate letters. $A_B^{G_i^n x_1 \ldots x_n}$ is the result of replacing each free occurrence of G_i^n with variables $t_1 \ldots t_n$ by $B_{t_1 \ldots t_n}^{x_1 \ldots x_n}$, providing that the following conditions are met:

(a) no subformula of A of the form $(v)C$ where v is a free variable of B other than $x_1 \ldots x_n$ contains a free occurrence of G_i^n

(b) for each t_j in an occurrence of G_i^n which is replaced, there is no subformula $(t_j)D$ in B which contains x_j.

If these conditions are not met, then $A_B^{G_i^n x_1 \ldots x_n}$ is A. Thus the axiom schema for instantiation will be: $(G_i^n)A \supset A_B^{G_i^n x_1 \ldots x_n}$. The rule of inference that from A you may infer $(G_i^n)A$ is sound.

EXERCISE 3. Prove that the axiom schemas are valid, but that if we omitted any of the restrictions in the definition of substitution they would not be.

EXERCISE 4. Show that the rule of inference is sound.

We will now consider N^2, the second order correlate of the number theory system N. The most important point is that we can take N^2 to consist of a single sentence because we can replace the axiom schema of induction with $(G_1^1)[G_1^1 0 \supset [(x)(G_1^1 x \supset G_1^1 x') \supset (x)G_1^1 x]]$. Thus N^2 will consist of the conjunction of Axioms 8 through 16 and the second order generalization of the induction schema.

CATEGORICITY OF N^2. *Any two models of N^2 are isomorphic.*

Proof. In order to make Axiom 8 true there must be distinct elements of the model assigned to $0, 0', 0'', \ldots$ Let S be the set of elements which are assigned to one of $0, 0', \ldots$ Since for any model $I(0) \in S$ and for all elements if $d_1 \in S$ then the unique element d_2 such that $\langle d_1, d_2 \rangle \in I(')$ will also be in S, we know that in order to satisfy the induction axiom every element of D must be in S. Thus in any model of N^2 every element of the domain is assigned to some $\underline{n}$. Given any two models of N^2 we can find the required correspondence by mapping the element of the first domain which is assigned to $\underline{n}$ onto the element of the second domain which is assigned to $\underline{n}$, for each $\underline{n}$.

COROLLARY 3. *For any sentence A of N^2, either $N^2 \vdash A$ or $N^2 \vdash -A$.*

COROLLARY 4. *For any sentence A of N^2, either $N^2 \supset A$ is valid or $N^2 \supset -A$ is valid.*

THEOREM. *There is no complete effective set of axioms and rules for second order logic.*

Proof. If A is a first order formula which is not valid then there is a model with domain consisting of the natural numbers in which A is false. If we let A' be the result of replacing the predicate constants F_i^n by the corresponding G_i^n, then if $\exists - A$ is the existential closure of $-A'$, i.e., the result of prefixing $(\exists G_i^n)$ for each free G_i^n in $-A'$, $\exists - A$ is true in any model with the natural numbers as domain. Thus A is invalid in first order logic iff $N^2 \supset \exists - A$ is valid in second order logic. If there were a complete effective axiomatization of second order logic we could find a recursive enumeration of the invalid formulas of first order logic. But we know this is impossible by Church's Theorem (p. 70).

THEOREM. *There is no complete arithmetic set of axioms and rules for second order logic.*

Proof. If there were a complete arithmetic set of axioms and rules then the proof relation would be arithmetically definable. Suppose that the relation were Σ_n or Π_n definable. We know that there is a Π_{n+1} set S which is not Σ_{n+1} or Π_n. But there is a formula B of N^2 such that $B\underline{m}$ is true iff $m \in S$. If there were a complete Σ_n or Π_n proof relation Bew_2 then $(Ez)\text{Bew}_2(z, `N^2 \supset B\underline{m}')$ iff $m \in S$ (by Corollary 4 above) and we would have a Σ_{n+1} definition of S.

EXERCISE 5. We will define the second order analogue of a

Henkin set to be a set Γ which satisfies conditions (a)–(g) (p. 5) and also condition (h):

(h) $(G_i^n)A$ is in Γ iff $A_{F_j^n}^{G_i^n}$ is in Γ for all F_j^n.

Show that there are sets which satisfy conditions (a)–(h) but which are not satisfiable.

In Chapter IX we proved that we could give a definition of truth for N in a theory whose vocabulary exceeded N only by containing one two-place relation not in the vocabulary of N. The theory consisted of a finite set of axioms characterizing the satisfaction relation and a definition of truth in terms of satisfaction. Let us take the added relational predicate to be F_1^2 and let $\theta(F_1^2)$ be the conjunction of the satisfaction axioms. What was proved then was that in the theory consisting of N plus $\theta(F_1^2)$, $\vdash (x)F_1^2x\,\text{`}A\text{'} \leftrightarrow A$, for every sentence A of N.

THEOREM. *The truth predicate for N can be defined in* N^2.

Proof 1. We define a formula Sxy of N^2 which expresses the satisfaction relation for N as $(G_1^2)(\theta(G_1^2) \supset G_1^2xy)$. Since satisfaction as defined by θ is a relation between numbers we know that there is some such relation among the set of ordered pairs of the domain of N^2 and this definition requires that Sxy holds only if G_1^2xy is true for all relations which satisfy θ. Truth of course will be defined as follows $T(\text{`}A\text{'}) =_{\text{df}} (x)Sx\,\text{`}A\text{'}$.

By the theorem in the previous chapter N, $\theta(F_1^2) \vdash (x)F_1^2x\,\text{`}A\text{'} \leftrightarrow A$, so $N^2 \vdash (G_1^2)[\theta(G_1^2) \supset ((x)G_1^2x\,\text{`}A\text{'} \leftrightarrow A)]$. Thus we know that both N^2, $\theta(G_1^2)$, $G_1^2x\,\text{`}A\text{'} \vdash A$ and $N^2$, $A$, $\theta(G_1^2) \vdash G_1^2x\,\text{`}A\text{'}$. From these it follows that $N^2 \vdash (x)[(G_1^2)(\theta(G_1^2) \supset G_1^2x\,\text{`}A\text{'})] \leftrightarrow A$. But the expression on the right of the biconditional is precisely $T(\text{`}A\text{'})$ and so the theorem is established.

EXERCISE 6. *Proof 2.* Define Sxy as $(\exists G_1^2)(\theta(G_1^2) \wedge G_1^2xy)$ and $T(\text{`}A\text{'})$ as $(\exists x)Sx\,\text{`}A\text{'}$ and prove the theorem.

COROLLARY 1. *$N^2 \vdash T$('A') iff A is a true sentence of N.*

Proof. If A is a true sentence of N then $N^2 \supset A$ and the "if" follows by the previous theorem. Assuming that no false sentences are provable in N gives the 'only if' for if $N^2 \vdash T$('A') then by the previous theorem $N^2 \vdash A$.

Having shown that the truth predicate for N is definable in N^2 in the strong sense just discussed, it is worth considering the possibility of defining truth for N^2 in N^2. The clauses of the truth definition would have to be formulated somewhat differently for if we want to treat sequences as first order objects then they cannot in the usual way assign sets to sets. Instead of treating sequences as functions and functional application as a relation between the function and a set, we could instead treat sequences as objects and functional application as a relation. That is, we will formulate our denotation relation Den(x, y, z) in such a way that if y is an individual variable or constant then for any sequence x there is a unique object z which stands in the relation to y. But if y is a predicate variable of constant then Den(x, y, z) will generally be true for numerous values of z.

If IndVar(x) is the formula which represents the predicate 'is an individual variable' and Pred$n$Var(x) is the predicate which represents 'is an n-place predicate variable', then our axioms for sequence existence and for characterizing Den would include:

$$(x)(y)[\text{Seq}(x) \wedge \text{IndVar}(y) \supset (Ez)(w)(\text{Den}(x, y, w) \leftrightarrow z = w)]$$

$$(x)[\text{Seq}(x) \supset (y)(w)[\text{IndVar}(y) \supset (Ez)\text{Seq}(z) \wedge z \underset{y}{\approx} x \wedge \text{Den}(z, y, w))]]$$

$$(x)[\text{Seq}(x) \supset (y)(G_1^1)[\text{Pred1Var}(y) \supset (Ez)(w)(\text{Seq})(z) \wedge z \underset{y}{\approx} x \supset (\text{Den}(z, y, w) \leftrightarrow G_1^1(w))]].$$

But when we want to give the analogous axiom for two-place predicate variables, we find that we need a four-place denotation relation, and in general for the n-place predicate variables we will need to use an $n+3$ place denotation relation. Thus no truth definition for N^2 can be given in the vocabulary of N^2.

Another question which our development of the truth theory might raise is whether we can find some analogue of the arithmetic hierarchy in second order arithmetic. The classification which we obtain by considering the types of quantified expressions which are required to define various predicates will be easier to deal with if, instead of formulating the hierarchy in second order logic as originally presented we consider an alternative formulation.

In Chapter IV we showed that for any formula of first order quantification theory there is an equivalent formula without quantifiers but instead containing some function symbols not in the original formula. The procedure for obtaining the equivalent formula was this: If A is the original formula let A_0 be the prenex normal form of A. Then A_1 will be the result of dropping any initial universal quantifiers and the first existential quantifier (Ey) and replacing all occurrences of y by $f(x_1 \ldots x_n)$ where $x_1, \ldots x_n$ are the free variables of A_0 whose universal quantifiers were dropped and f is a function symbol not in A_0. A_2 is obtained by applying this process to A_1 and so on until a quantifier free formula is obtained. (At each step we choose a new function symbol f.) The system we will now consider is the one we obtain by allowing quantification of the function symbols in the quantifier free system.

We will call the system Skolem second order logic. The primitive vocabulary will be the same as that of second order logic as presented in this chapter except that in place of the various types of predicate variables G_i^n and constants F_i^n we will have function variables g_i^n and constants f_i^n. The model theory will be the natural modification of the previous model theory, namely we will require that sequences assign to n-place function variables functions from D^n to D. A sequence α will satisfy $(g^n)A$ iff every sequence $\beta \underset{g^n}{\approx} \alpha$ satisfies A.

We can show that for any formula of second order logic there is a translation into Skolem second order logic such that the first formula is satisfiable in $\langle D, I\rangle$ iff the second is. There is also a converse translation which has the same property. To translate from second order logic into Skolem logic we replace each successive well formed part of A of the form $(G_i^n)B$ by

$$(g_i^n)(z)B^{G_i^n(x_1 \ldots x_n)}_{g_i^n(x_1 \ldots x_n)=z}.$$

Clearly this process preserves satisfiability since for each set S of n-tuples of D^n which could be assigned to G_i^n for every element d of D there is a function such that $g(x_1, \ldots x_n) = d$ iff $\langle x_1 \ldots x_n \rangle \in S$, and conversely each choice of function and element will determine a set of n-tuples which could be assigned to G_i^n. To obtain the converse mapping we replace each part of the form

$$(g_i^n)B \text{ by } (G_i^{n+1})[(x_1)\ldots(x_n)(Ez)(y)(G_i^{n+1}(x_1, \ldots x_n, y) \leftrightarrow$$
$$\leftrightarrow y = z) \supset B^{g_i^n(x_1, \ldots x_n)=y}_{G_i^{n+1}(x_1, \ldots x_n, y)}].$$

In this case the translation will be vacuously satisfied for values of G_i^{n+1} which do not determine functions, so whether the quantified formula will be satisfied depends on whether B holds for all functions which are assigned to G_i^{n+1}.

EXERCISE 7. Show that if we take a formula A of Skolem logic and translate it into a formula B of second order logic and then translate B into a formula C in Skolem logic that $\models A \leftrightarrow C$.

We will now show that every formula of Skolem second order arithmetic which consists of a string of quantifiers prefixed to a recursive predicate is equivalent to a formula

$$(QV_1)(QV_2)\ldots(QV_n)(Qv)R,$$

where all of the V_i are function variables, R is recursive and adjacent quantifiers alternate between existential and universal. Such a formula we will say is in *analytic form*. If a predicate $B(n)$ can be written in analytic form with n function quantifiers beginning with an existential quantifier it is said to be Π_n^1, if it can be written in analytic form with n quantifiers where the initial quantifier is universal it is Σ_n^1. Predicates which are both Π_n^1 and Σ_n^1 are Δ_n^1. A predicate is analytic if it can be written in one of these forms for some n.

THEOREM. *Every predicate expressible in N^2 is analytic.*

Proof. We need only show that we can always find an equivalent expression in analytic form, which will follow from several lemmas:

LEMMA 1. *Every formula of N^2 is equivalent to a formula in prenex normal form.*

Proof. As in first order logic by moving quantifiers out.

LEMMA 2. *$(x)B(x)$ is equivalent to $(g_i^1)B(g_i^1(0))$, if g_i^1 does not occur in B, is the first function variable not in $B(x)$. $(Ex)B(x)$ is equivalent to $(Eg_i^1)B(g_i^1(0))$, if g_i^1 does not occur in $B(x)$.*

LEMMA 3. *The following equivalences hold*:

$$(x)(Eq)B \leftrightarrow (Eq)(x)B \qquad (Ex)(g)B \leftrightarrow (g)(Ex)B.$$

Proof. EXERCISE 8.

LEMMA 4. *The following equivalences hold*:

$$(g_i^n)(g_j^m)B(g_i^n(x_1, \ldots x_n), g_j^m(y_1, \ldots y_m)) \leftrightarrow$$
$$\leftrightarrow (g_k^{n+m})B((g_k^{n+m}(x_1, \ldots x_n, y_1, \ldots y_m)_1,$$
$$(g_k^{n+m}(x_1 \ldots x_n y_1 \ldots y_m))_2)$$

$$(Eg_i^n)(Eg_j^m)B(g_i^n(x_1, \ldots x_n), g_j^m(y_1, \ldots y_m)) \leftrightarrow$$
$$\leftrightarrow (Eg_k^{n+m})B(g_k^{n+m}(x_1, \ldots x_n, y_1, \ldots y_m)_1, (g_k^{n+m}(x_1 \ldots x_n y_1 \ldots y_m))_2).$$

Proof. Suppose the right hand side of the first biconditional is true, then among the values of g_k^{n+m} are $2^{g_i^n}(x_1, \ldots x_n)3^{g_j^m}(y_1, \ldots y_m)$ for all values of g_i^n and g_j^m, so the left hand side must be true. If the right hand side is false for some value of $x_1, \ldots x_n, y_1, \ldots y_m$ let

$$g_i^n(x_1, \ldots x_n) = (g_k^{n+m}(x_1, \ldots x_n, y_1, \ldots y_m))_1$$

and

$$g_j^m(y_1, \ldots y_m) = (g_k^{n+m}(x_1, \ldots x_n, y_1, \ldots y_m))_2$$

and the left hand side will be false for those values. The second biconditional is similar.

Now we can show that any prenex formula in Skolem N^2 can be written in analytic form. First, we can move all quantifiers of individual variables to the extreme right. Next we can replace all adjacent strings of universal or existential quantifiers by single quantifiers of that type. If the remaining individual quantifiers are more than one in number we replace them by function quantifiers by Lemma 2. If there is only one individual quantifier and it is of the same type as the last function quantifier we replace the individual variable by a function variable and then replace the resulting adjacent similar quantifiers by a single quantifier by Lemma 4. Finally, if there is no individual variable in the innermost position, we insert a vacuous one.

The methods used in Chapter VII can be used to show that the classification of predicates does produce a hierarchy, i.e., that each category Σ_n^1 and Π_n^1 does properly include lower categories. The predicates of Δ_0^1 form are said to be hyperarithmetic. When the hierarchy was first defined it was expected that the hyperarithmetic sets would coincide with the arithmetic sets but this has been proved false.

EXERCISE 9. Show that the hyperarithmetic sets *include* the arithmetic sets. (Use Exercise 6.)

The characterization of the analytic hierarchy could have been given in terms of our original second order logic rather than Skolem logic, but it would have been rather more cumbersome. We will now give another example of the usefulness of the Skolem form of second order logic. It has been recently suggested that first order logic is inadequate to express certain statements of natural languages. For example, if there are four place 'atomic' expressions of English $F(x, y, z, w)$ then it might be the case that for every value of x we can find a value of y and for every value of z we can find a w such that $F(x, y, z, w)$. If the choice of y depends only on x and the choice of w depends only on z, then neither $(x)(Ey)(z)(Ew)Fxyzw$ nor $(z)(Ew)(x)(Ey)Fxyzw$ fully captures the truth of the matter for the first of these is true even if the choice of a value of w depends on x and y as well as z. Similarly the second is true even if the choice of y depends on z and w as well as x.

We are not concerned here with the question whether such examples actually arise in ordinary everyday language, but only with the fact that such situations are perfectly possible mathematically. For example, if we added to N an atomic predicate interpreted as $x < y \wedge$

$w = z^2$, we would have such a case. One suggestion for formalizing such statements has been that we permit branching quantifiers. That is, if $(Qx_1)\ldots(Qx_n)$ are strings of quantifiers and B is a formula, then

$$\begin{matrix} (Qx_1)\ldots(Qx_n) \\ (Qx_{n+1})\ldots(Qx_m) \\ \vdots \\ (Qx_{m+1})\ldots(Qx_{m+k}) \end{matrix} \quad B(x_1 \ldots x_{m+n})$$

is also well formed.

For example, the case we were considering above could be written as $\begin{smallmatrix}(x)(Ey)\\(z)(Ew)\end{smallmatrix}Fxyzw$. The question we want to settle is what relation this type of logic bears to first and second order logic. Using the Skolem logic we can answer these questions rather easily. We note first that any formula of branching quantifier theory will be satisfied in a model iff there are functions $f_1, \ldots f_n$ which satisfy the Skolem function form of the formula. The Skolem function form is obtained as it was in the case of first order logic, with the stipulation that we treat each branch of quantifiers independently.

That is, when we replace an existentially quantified variable by a function symbol, the only variables which are arguments of that function are the variables which are bound by universal quantifiers which precede the existential quantifier in that branch of quantifiers. In the case we were considering before, the Skolem function form would be $F(x, f_1(x), z, f_2(z))$. A formula in branching quantifier logic is valid in a model iff there are functions $f_1, \ldots f_n$ which satisfy the Skolem formula in that model. Thus a formula in branching quantifier logic is valid iff the existential quantification $(Eg_1)\ldots(Eg_n)B$ of its Skolem form is valid. We will say that a formula in this form, i.e. all of whose quantifiers are initial existential quantifiers of function variables is in purely existential Skolem form.

EXERCISE 10. Show that there is a formula in purely existential Skolem form which is equivalent to $(\infty x)Gx$.

EXERCISE 11. Show that the set of valid formulas in purely existential Skolem form is not Σ_1.

EXERCISE 12. Show that if A is in purely existential Skolem form and A is satisfiable, then A is satisfiable in a finite or denumerably infinite model.

Exercises 10–11 show that branching quantifier logic is not definable within first order logic; Exercise 12 shows that branching quantifier logic is equivalent to a proper subpart of second order logic since every formula in branching quantifier logic is equivalent to some second order formula but not conversely.

CHAPTER XIII

ALGEBRAIC LOGIC

In this chapter we will present alternative formulations of first order logic, formulations which are intended to make more perspicuous the connections between syntactic structures and semantic operations. In the discussion of model theory in earlier chapters we defined a relation of satisfaction which holds between a model, a formula and a sequence of elements from the domain of the model. Given that definition we can associate with each formula A and interpretation I a set of sequences $I[A] = \{\alpha: \alpha \text{ satisfies } A \text{ in } I\}$.

EXAMPLE 1. Prove that A and B are logically equivalent iff $I[A] = I[B]$ for all I.

If we let D^ω denote the set of all sequences formed from the set D, then we note that A is true in I iff $I[A] = D^\omega$, where D is the domain of I. Another useful piece of notation will be $D^\omega - S$ to stand for the set of sequences formed from D which are not in S. With these conventions we can show that $I[A]$ for truth functional A is a simple function of the value for the components. In particular,

$$I[A \wedge B] = I[A] \cap I[B], \qquad I[A \vee B] = I[A] \cup I[B],$$
$$I[-A] = D^\omega - I[A],\ I[A \supset B] = (D^\omega - I[A]) \cup I[B].$$

This suggests that one could equally naturally give a direct definition of $I[A]$ without going through the definition of satisfaction. In order to do this for quantificational formulas we must introduce two further operations on sets of sequences C_j and U_j, which are called cylindrification and universalization respectively.

$$C_j(S) = \{\alpha: \alpha \in D^\omega \text{ and } (\exists\beta)\beta \underset{v_j}{\approx} \alpha \text{ and } \beta \in S\}$$
$$U_j(S) = \{\alpha: \alpha \in D^\omega \text{ and } (\beta) \text{ if } \beta \underset{v_j}{\approx} \alpha \text{ then } \beta \in S\}.$$

EXAMPLE 2. Show that if $S \subseteq D^\omega$ then $U_j(S) = D^\omega - C_j(D^\omega - S)$.

Cylindrification is so-called because of the following special case. If we consider three element sequences from the domain of real numbers then each sequence corresponds to a spatial point. The cylindrification of a set of such sequences will form a sequence which is an infinite figure in space. More specifically, if we take for example the points $\{\langle 0, x, y\rangle: x^2 + y^2 = k\}$ as our set S, then C_1S will be a cylinder around the first axis.

With the operations we defined, we can characterize $I[A]$ directly by clauses parallel to those for satisfaction: If A is atomic, i.e., $F^n v_{i_1} \ldots v_{i_n}$, then $I[A] = \{\alpha : \langle \alpha(i_1) \ldots \alpha(i_n)\rangle \in I(F^n)\}$

$$I[-A] = D^\omega - I[A]$$
$$I[A \vee B] = I[A] \cup I[B]$$
$$I[A \wedge B] = I[A] \cap I[B]$$
$$I[A \supset B] = (D^\omega - I[A]) \cup I[B]$$
$$I[(Ex_j)A] = C_j(I[A])$$
$$I[(x_j)A] = U_j(I[A]).$$

EXAMPLE 3. Prove that if A is a sentence, then $I[A]$ is either D^ω or the empty set.

Once we have formulated our semantic theory in this way, it is tempting to reconsider the syntax of logic and reformulate it so as to reflect the semantics more closely. In the cases of the sentential connectives there is already a close parallel but for the quantifiers the syntax could be simpler. In other words, we might consider adopting a new language with symbols for the operations of cylindrification and universalization instead of quantifiers. One of the benefits of such a formulation would be that it appears that we can avoid the use of variables entirely since the cylindrification and universalization operations mention only indices of variables. However, if we are to avoid variables completely we must also pay attention to the definition of $I[A]$ for atomic A. Note that this definition depends on the variables which are present. For example, $I[F^2x_1x_2]$ is different in general from $I[F^2x_3x_4]$.

One possibility which suggests itself is that we could change the definition of an interpretation so that $I(F^2)$ would be a set of infinite sequences. In the new sense of interpretation, $I(F^2)$ would be what $I[F^2x_1x_2]$ was in the old sense. In order to use this definition, however, it would also be necessary to make some provision for constructing the set of sequences $I[F^2x_nx_m]$ where $n \neq 1$ or $m \neq 2$. One way of making provision for these formulas is to treat them as defined in terms of F^2, identity and the other logical operations. That is we can make use of the fact that $F^2x_nx_m$ is equivalent to $(Ex_1)(Ex_2)(F^2x_1x_2 \wedge x_1 = x_n \wedge x_m = x_2)$. This reduces the general problem of using variables other than the initial ones in atomic formulas to the specific problem of using variables in identity formulas. Fortunately, identity is a logical operation and we can characterize in general the expressive power of identity in terms of operations on sets of sequences. The effect of $x_i = x_j$ is to form the set of sequences such that the ith and jth elements are identical. Thus we will add to the semantics the expressions $D_{i,j}$ for each i and j. (These operations are called diagonalization because of a geometric analogy similar to the one for cylindrification. If we consider three element sequences, i.e., areas in a three dimensional space, then $D_{1,2}$, $D_{2,3}$ and $D_{1,3}$ are all planes diagonal to the coordinates. Hence the use of D.)

$$I[D_{i,j}] = \{\alpha\colon \alpha \in D^\omega \text{ and } \alpha(i) = \alpha(j)\}.$$

In formulating our new language which we will call Cylindrification Theory (CT) we will have an expression corresponding to each operation in the semantics. We will use the same letter underlined for the syntactic expression of the operation. Thus C is the operation of cylindrification and $\underline{C}$ is the expression of the object language. We will simplify our language slightly by making use of the fact that universalization is definable in terms of cylindrification and complementation, and the fact that the conditional and biconditional are definable in terms of the other sentential connectives. Since we are construing our formulas as standing for sets of sequences in any given interpretation it is fairly natural to formulate the theory as consisting of equations between formulas. We will also add a constant formula $\underline{T}$ which is to stand for the universal set of sequences. Thus the official vocabulary of CT will consist of:

An infinite list of predicate letters F_i^n for each i and n.

$$-, \wedge, \vee, \underline{C}_i, \underline{D}_{i,j}, =, \underline{T}, \text{ for each } i \text{ and } j.$$

An *atomic CT formula* is any predicate letter, any $\underline{D}_{i,j}$, or $\underline{T}$. A is a *CT formula* iff A is an atomic CT formula or A is $-B$, $B \wedge E$, $B \vee E$, or $\underline{C}_i B$ where B and E are CT formulas. A *CT equation* is any string of the form $A = B$ where A and B are CT formulas.

Before stating the definition of an interpretation of the CT language, it will be useful to add one more piece of terminology. We will say that a set of sequences $S \subseteq D^\omega$ is *uniform beyond n* if for all $m > n$, $C_m S = S$. Less formally, to say that a set of sequences is uniform beyond n is to say that in order to determine whether a sequence belongs to the set it is sufficient to know the initial segment of the sequence to up the nth term. It is a characteristic feature of the definition of interpretation and of the sets of sequences assigned to formulas in this language that all sets of sequences involved are uniform beyond n for some n.

An interpretation of CT is an ordered pair $\langle D, I\rangle$ such that D is a non-empty set and I is a function such that for all F_i^n, $I(F_i^n)$ is a set of infinite sequences from D which is uniform beyond n. Next we define $I[A]$, the set of sequences assigned to A by I for formulas in general.

$$I[F_i^n] = I(F_i^n)$$
$$I[A \wedge B] = I[A] \cap I[B]$$
$$I[-A] = D^\omega - I[A]$$
$$I[A \vee B] = I[A] \cup I[B]$$
$$I[\underline{D}_{i,j}] = \{\alpha : \alpha \in D^\omega \text{ and } \alpha(i) = \alpha(j)\}$$
$$I[\underline{C}_i A] = C_i I[A]$$
$$I[\underline{T}] = D^\omega.$$

An equation $A = B$ is true in $\langle D, I\rangle$ iff $I[A] = I[B]$. You should note that in general it is not the case that, if $A = B$ is not true, then $-A = B$ is true. The symbol $-$ is an operation on the formula A and does *not* negate the equation in $-A = B$.

Having defined the syntax and semantics for CT we can now state and prove the relation between CT and quantification theory. We will consider a quantificational language based on negation, conjunction, disjunction and existential quantification in order to facilitate the comparison. We will define a one-one translation between formulas of CT and a subset of the formulas of quantification theory, namely those formulas in standard variable form. An atomic formula is in *standard variable form* if it is an identity or if it is F_i^n followed by the variables $x_1, \ldots x_n$ in that order. Thus $F^3x_1x_2x_3$ is in standard variable form (svf) but $F^3x_1x_3x_2$, $F^3x_2x_2x_4$, $F^3x_1x_2x_2$ and $F^3x_3x_2x_1$ are not. A formula is in svf iff every atomic formula in it is in svf. We will now define our translation between quantificational formulas in svf and formulas of CT:

$\not{c}(F_i^n x_1 \ldots x_n)$ will be $\underline{F}_i^n$

$\not{c}(x_i = x_j)$ will be $\underline{D}_{i,j}$

$\not{c}(-A)$ will be $-(\not{c}A)$

$\not{c}(A \wedge B)$ will be $(\not{c}A) \wedge (\not{c}B)$

$\not{c}(A \vee B)$ will be $(\not{c}A) \vee (\not{c}B)$

$\not{c}((\exists x_i)A)$ will be $\underline{C}_i(\not{c}A)$.

This translation specifies a formula in CT corresponding to every formula in svf in quantification theory. Taken in reverse, it also specifies for all formulas of CT not containing $\underline{T}$, a formula of quantification theory. Thus if we add that $\underline{T}$ is translated as $x_1 = x_1$, we have a complete translation of the formulas of CT into quantification theory; we will call this translation $\not{c}^*$.

Next we need to give a method for transforming interpretations of the quantification language into interpretations of the cylindrification theory and vice versa. In presenting these transformations it will be useful to have the *uniform infinite extension operator* UIO. $\text{UIO}(D, S) = \{\alpha: \alpha \in D^\omega$ and for some n, $\langle \alpha(1), \ldots \alpha(n) \rangle \in S$. Thus UIO takes a set S and forms all of the infinite sequences of elements of D such that some initial segment of that sequence is in S. (Where context makes it clear what the domain is, we will often omit reference to the domain and simply write UIO(S).) Now for each interpretation $\langle D, I \rangle$ of quantification theory we can define an interpretation $\langle D, I\not{c} \rangle$ of CT by specifying that $I\not{c}(F_i^n) = \text{UIO}(D, I(F_i^n))$.

Conversely, for each interpretation $\langle D, I\rangle$ of CT we can define an interpretation $\langle D, I^*\rangle$ of quantification theory by requiring that $I^*(F_i^n) = D^n \cap I(F_i^n)$.

THEOREM 1. *If A is a quantificational formula in svf, and* $\langle D, I\rangle$ *is an interpretation of the quantificational language, then* $\{\alpha\colon \alpha$ *satisfies* A *in* $\langle D, I\rangle\} = I\not{c}[\not{c}(A)]$; *if A is a formula of* CT *and* $\langle D, I\rangle$ *is an interpretation of* CT *then* $I[A] = \{\alpha\colon \alpha$ *satisfies* $\not{c}^*(A)$ in $\langle D, I^*\rangle\}$.

EXERCISE 4. Prove Theorem 1. (Use induction on the order of formulas.)

COROLLARY. *A is valid in quantification theory iff* $\not{c}(A) = \underline{T}$ *is true in every* CT *interpretation.*

Proof. If A is not valid, then by the theorem there is a CT interpretation $\langle D, I\rangle$ such that $I[\not{c}(A)] \neq D^\omega$, and thus $\not{c}(A) = \underline{T}$ is false in that interpretation. If there is an interpretation $\langle D, I\rangle$ in which $\not{c}(A)$ is not assigned D^ω, then there is a quantificational interpretation $\langle D, I^*\rangle$ such that some sequence fails to satisfy $\not{c}^*(\not{c}(A))$, but $\not{c}^*(\not{c}(A))$ is A.

In order to extend the relation between CT formulas and quantificational formulas in general, it will suffice to prove the following:

LEMMA. *For any quantificational formula, there is an equivalent formula in svf.*

EXERCISE 5. Prove this lemma. (Use induction on the number of variables not in standard form.)

One of the advantages of this alternative conception of logic is that we can present an axiomatization in the form of a system of equations. The axioms of CT are the following:

(1) $A \wedge B = B \wedge A$
(2) $A \vee B = B \vee A$
(3) $A \wedge (B \vee H) = (A \wedge B) \vee (A \wedge H)$
(4) $A \vee (B \wedge H) = (A \vee B) \wedge (A \vee H)$

(5) $A \vee \perp = A$

(6) $A \wedge T = A$

(7) $A \vee -A = \underline{T}$

(8) $A \wedge -A = \perp$

(9) $\underline{C}_j \underline{\perp} = \underline{\perp}$

(10) $A \vee \underline{C}_j A = \underline{C}_j A$

(11) $\underline{C}_j(A \wedge \underline{C}_j B) = \underline{C}_j A \wedge \underline{C}_j B$

(12) $\underline{C}_j \underline{C}_k A = \underline{C}_k \underline{C}_j A$

(13) $\underline{D}_{kk} = \underline{T}$

(14) $\underline{D}_{k,j} = \underline{C}_m(\underline{D}_{k,m} \wedge \underline{D}_{m,j})$ if $m \neq j$, $m \neq k$

(15) $\underline{C}_k(\underline{D}_{k,j} \wedge A) \wedge \underline{C}_{k,j} \wedge -A) = \underline{\perp}$ if $j \neq k$.

$\perp \underset{\text{df}}{=} -T$. The only rule is the substitution of equalities. A proof is a sequence of equations $E_1, \ldots E_k$ such that each equation is either an instance of an axiom or follows from previous equations by substitution of equality. We will write $\vdash A = B$ to mean that $A = B$ is provable, which is the case when there is a proof whose last equation is $A = B$.

We will illustrate the methods of proof by showing the distribution of cylindrification over disjunction, i.e., $\underline{C}_j(A \vee B) = \underline{C}_j A \vee \underline{C}_j B$. We first prove some general results which are of use in proving equations. If $A \wedge -B = \underline{\perp}$, then $A \wedge B = B$. By the assumption and substitution, $(A \wedge -B) \vee B = \underline{\perp} \vee B$, so by axiom schemata (3) and (5),

$$(A \wedge B) \vee (B \wedge -B) = B, \text{ so by schema (8)},$$

$$(A \vee B) \vee \underline{\perp} = B \text{ and by (5) we obtain } A \wedge B = B.$$

We can now establish a very useful derived rule of indirect proof: If $\vdash A \wedge -B = \perp$ and $\vdash -A \wedge B = \perp$, then $\vdash A = B$. This follows easily from the previous fact since the assumptions of the derived rule give us that $\vdash A \wedge B = B$ and $\vdash A \wedge B = A$, whence $\vdash A = B$.

EXERCISE 6. Prove the following equations: $(A \vee B) \vee C = A \vee (B \vee C)$, $(A \wedge B) \wedge C = A \wedge (B \wedge C)$, $-(A \vee B) = -A \wedge -B$, $-(A \wedge B) = -A \vee -B$, $A \vee A = A$, $A = A$, $A \vee T = T$.

Next we prove some facts about cylindrification, mainly that repeated applications of the same cylindrification produce nothing new.

(A) $\underline{C}_jT = T$. *Proof.* By axiom schema (10), $T \vee \underline{C}_jT = \underline{C}_jT$ and by the exercises $T \vee \underline{C}_jT = T$.

(B) $\underline{C}_j\underline{C}_jA = \underline{C}_jA$. By using instances of axiom schemata (1) and (6), we prove that $\underline{C}_j\underline{C}_jA = \underline{C}_j(T \wedge \underline{C}_jA)$, so by schema (11), $\underline{C}_j\underline{C}_jA = \underline{C}_jT \wedge \underline{C}_jA$, so by (A) we obtain $\underline{C}_j\underline{C}_jA = \underline{C}_jA$.

(C) $\underline{C}_j - \underline{C}_jA = -\underline{C}_jA$. We will use the method of indirect proof established above.

$$\begin{aligned}
\underline{C}_j - \underline{C}_jA \wedge --\underline{C}_jA &= \underline{C}_j - \underline{C}_jA \wedge \underline{C}_jA \quad \text{by the exercises.}\\
&= \underline{C}_j - \underline{C}_jA \wedge \underline{C}_j\underline{C}_jA \quad \text{by (A).}\\
&= \underline{C}_j(\underline{C}_j - \underline{C}_jA \wedge \underline{C}_jA) \quad \text{by schema (1) and (11).}\\
&= \underline{C}_j\underline{C}_j(-\underline{C}_jA \wedge \underline{C}_jA) \quad \text{by schema (11) again.}\\
&= \underline{C}_j\underline{C}_j\bot \quad \text{by schema (8).}\\
&= \bot \quad \text{by two applications of schema (9).}\\
-\underline{C}_j - \underline{C}_jA \wedge -\underline{C}_jA &= -(\underline{C}_j - \underline{C}_jA \vee \underline{C}_jA) \quad \text{by the exercises.}\\
&= -((\underline{C}_j - \underline{C}_jA \vee -\underline{C}_jA) \vee \underline{C}_jA) \quad \text{by schema (10).}\\
&= -T \quad \text{by schema (7) and the exercises.}\\
&= \bot \quad \text{by definition.}
\end{aligned}$$

Thus we have proved that $\underline{C}_j - \underline{C}_jA \wedge --\underline{C}_jA = \bot = -\underline{C}_j - \underline{C}_jA \wedge -\underline{C}_jA$, so by the method of indirect proof $\underline{C}_j - \underline{C}_jA = -\underline{C}_jA$.

(D) $\underline{C}_j(A \vee B) = \underline{C}_jA \vee \underline{C}_jB$.

Proof. We again use the indirect method:

$$\begin{aligned}
\underline{C}_j(A \vee B) \wedge -\underline{C}_jA \wedge -\underline{C}_jB &= \underline{C}_j(A \vee B) \wedge \underline{C}_j -\underline{C}_jA \wedge \underline{C}_j - \underline{C}_jB \quad \text{by (B).}\\
&= \underline{C}_j((A \vee B) \wedge \underline{C}_j - \underline{C}_jA \wedge \underline{C}_j \underline{C}_jB) \quad \text{by schema (11).}\\
&= \underline{C}_j((A \vee B) \wedge -\underline{C}_jA \wedge -\underline{C}_jB) \quad \text{by (B) again.}
\end{aligned}$$

$= \underline{C}_j((A \vee B) \wedge -(\underline{C}_jA \vee A) \wedge$
$-(\underline{C}_jB \vee B))$ by schema (10).
$= \underline{C}_j((A \vee B) \wedge -\underline{C}_jA \wedge -A \wedge$
$-\underline{C}_jB \wedge -B)$ by the exercises.
$= \underline{C}_j\bot$ by schema (8) and the exercises.
$= \bot.$

$-\underline{C}_j(A \vee B) \wedge (\underline{C}_jA \vee \underline{C}_jB) = (-\underline{C}_j(A \vee B) \wedge \underline{C}_jA) \vee$
$(-\underline{C}_j(A \vee B) \wedge \underline{C}_jB$
by schema (3).
$= (\underline{C}_j - \underline{C}_j(A \vee B) \wedge \underline{C}_jA) \vee$
$(\underline{C}_j - \underline{C}_j(A \vee B) \wedge \underline{C}_jB)$
by (C).
$= \underline{C}_j(\underline{C}_j - \underline{C}_j(A \vee B) \wedge A) \vee$
$\underline{C}_j(\underline{C}_j - \underline{C}_j(A \vee B) \wedge B)$ by 11.
$= \underline{C}_j(-\underline{C}_j(A \vee B) \wedge A) \vee \underline{C}_j(-\underline{C}_j$
$(A \vee B) \wedge B)$ by (C) again.
$= \underline{C}_j(-(\underline{C}_j(A \vee B) \vee (A \vee B)) \wedge A)$
$\vee\, \underline{C}_j(-(\underline{C}_j(A \vee B) \vee (A \vee B))$
$\wedge B)$ by schema (10).
$= \underline{C}_j(-\underline{C}_j(A \vee B) \wedge -A \wedge -B \wedge$
$A) \vee \underline{C}_j(-\underline{C}_j(A \vee B) \wedge -A \wedge$
$-B \wedge B).$
$= \underline{C}_j\bot \vee \underline{C}_j\bot.$
$= \bot.$

EXERCISE 7. Prove $-\underline{C}_j(A \wedge -\underline{C}_jB) = -(\underline{C}_jA \wedge -\underline{C}_jB)$.

The theory of cylindrification presents a very direct connection between the syntax and semantics of the language and clarifies in some respects the concept of quantification. Two features of the theory, however, are not quite as elegant as we would wish. For any formula of the language there is an n such that the set of sequences associated with the formula in any interpretation is uniform after n. In other words, although we assign sets of infinite sequences to the formulas, in fact, whether or not a sequence is assigned to a formula depends only on some finite number of arguments. This suggests that we might seek a similar theory in which the sequences are finite. Secondly, the infinite set of operations C_i and the set $D_{i,j}$ seem intuitively to be many minor variations on two basic operations. Thus we might also be led to consider whether we can formulate a theory in which we have only a finite number of intuitively distinct operations.

The most obvious way to finitize the theory is to use the operations C_1 and $D_{1,2}$ and to define all of the other C_j and $D_{i,j}$ by using the first two and permutation operators. That is, to define C_3 we would first find an operator which turned a set of sequences $\langle a_1, \ldots a_n \rangle$ into the corresponding set of sequences $\langle a_3 a_1 a_2, \ldots a_n \rangle$, apply C_1 and then apply the operator which would reverse the work of the first permutation. You should note that whereas in CT the formula $\underline{D}_{i,j}$ denoted the set of all infinite sequences in the interpretation in which the ith and jth elements were identical we will have a multiplicity of $\underline{D}_{i,j}$ formulas in the new language. For each n, we will need to have the set of n-tuples of the domain in which the ith and jth elements are the same.

For this and other reasons we will need to add an operator $\#$ which will lengthen sequences by one place. Another of the reasons for having this operator is that if we wish to take the conjunction of a 1-place and a 2-place predicate, e.g., $F^1 \wedge G^2$ we will always obtain the empty set (since it is the intersection of a set of 1-tuples with a set of pairs, which is always empty) unless we first fatten the set of unit sequences to form a set of pairs.

Since we are working with finite sequences we can also consider an alternative to cylindrification as the operation analogous to existential quantification. In the cylindrification of a set of sequences we form the set of all sequences which are like one of the initial sequences except possibly at the cylindrified argument. An alternative would be to eliminate the argument place which is being operated on and to form the set of sequences such that some way of filling in the extra argument would give a sequence in the original set. We will illustrate the difference between this operator E and C in a simple case before stating the general definition. Suppose that our domain consists of

$\{0, 1, 2\}$ and that we are considering sequences of length 2. Furthermore, let F^2 be assigned $\{\langle 0, 1\rangle, \langle 1, 2\rangle\}$. Then $\underline{C}_2F^2$ would be assigned the sequences $\{\langle 0, 1\rangle, \langle 0, 2\rangle, \langle 0, 0\rangle, \langle 1, 0\rangle, \langle 1, 1\rangle, \langle 1, 2\rangle\}$ whereas $\underline{E}_2F^2$ would form the set of sequences $\{\langle 0\rangle, \langle 1\rangle\}$. In CT there was no point to using E rather than C since the result of eliminating one place from an infinite sequence is still an infinite sequence. However, since we are now working with finite sequences there is a point to using E since we wish to preserve in the formal system the feature that the length of the sequence assigned to a formula reflects the number of argument places which are relevant in the formula. Thus we would like formulas corresponding to closed sentences to be assigned null sequences of elements. This will be the case for $\underline{E}_1\underline{E}_2F^2$ for example but would not be for $\underline{C}_1\underline{C}_2F^2$. (There is of course a simple and systematic relation between $\underline{C}$ and $\underline{E}$, so that for example $\underline{E}_1\underline{E}_2F^2$ will either be assigned the null sequence, in which case $\underline{C}_1\underline{C}_2F^2$ will be assigned D^2, or both would be assigned the empty set of sequences.) As we will show later C can be defined in terms of E and our other operations.

Thus our basic semantic operations will be $-$, $\wedge$, $\vee$, E, D, P, R, Я, and $\#$; P, R and Я being the permutation operators to be defined. The corresponding syntactic names for the operations will be constructed as before: The language ET consists of an infinite list of predicate letters F_i^n where i, n range over non-negative integers, $\underline{T}$, $-$, $\wedge$, $\vee$, $\underline{E}$, $\underline{D}$, $\underline{P}$, $\underline{R}$, $\underline{\text{Я}}$, $\underline{\#}$ and $=$, (,). Every formula of the language will be an n-formula for some non-negative n; the point of categorizing formulas in this way will become clear when we state the semantics and show that an n-formula is always assigned a set of n-tuples.

$\underline{T}$ is a 0-formula. $\underline{D}$ is a 2-formula. F_i^n is an n-formula for all i. If A and B are n-formulas, then $-A$, $A \wedge B$, $A \vee B$, $\underline{P}A$, $\underline{R}A$, $\underline{\text{Я}}A$ are n-formulas also. If A is an $n+1$ formula, then EA is an n-formula and if A is an n-formula $\underline{\#}$ is an $n+1$ formula. A string of symbols is a *formula* iff it is an n-formula for some n. An equation is any string of the form $A = B$ where A and B are formulas.

We will now give precise definitions of the logical operations and then of the concept of an interpretation for this language. We will use

σ as a variable over sequences of length n, and $\sigma(i)$ to indicate the ith element of σ.

$$E(S) = \{\langle\sigma(2), \sigma(3), \ldots \sigma(n)\rangle : \sigma \in S\}$$
$$D(S) = \{\sigma : \sigma \in S \text{ and } \sigma(1) = \sigma(2)\}$$
$$P(S) = \{\langle\sigma(2), \sigma(1), \sigma(3) \ldots \sigma(n)\rangle : \sigma \in S\}$$
$$R(S) = \{\langle\sigma(2), \sigma(3), \ldots \sigma(n), \sigma(1)\rangle : \sigma \in S\}$$
$$\mathcal{R}(S) = \{\langle\sigma(n), \sigma(1), \ldots \sigma(n-1) : \sigma \in S\}$$
$$\#(S) = \{\langle d, \sigma(1), \sigma(2) \ldots \sigma(n)\rangle : d \in D \text{ and } \sigma \in S\}.$$

An interpretation of the language ET will be an order pair $\langle D, I\rangle$ where D is a non-empty set and where I is a function such that $I(F_i^n) \subseteq D^n$. We can now define $I[A]$ analogously to the earlier definition.

$$I[F_i^n] = I(F_i^n) \quad I[\underline{T}] = \{\langle\ \rangle\}$$
$$I[\underline{D}] = \{\langle d, d\rangle : d \in D\}$$
$$I[A \vee B] = I[A] \cup I[B]$$
$$I[A \wedge B] = I[A] \cap I[B]$$
$$I[-A] = D^n - I[A], \text{ where } A \text{ is an } n\text{-formula.}$$
$$I[\underline{E}A] = E(I[A])$$
$$I[\underline{P}A] = P(I[A])$$
$$I[\underline{R}A] = R(I[A])$$
$$I[\underline{\mathcal{R}}A] = \mathcal{R}(I[A])$$
$$I[\underline{\#}A] = \#(I[A]).$$

LEMMA 1. *Every n-formula A is assigned a set of n-tuples $I[A]$.*

Proof. By induction on the order of the n-formulas (not on n).

An equation $A = B$ is true in a model $\langle D, I\rangle$ iff $I[A] = I[B]$. An equation is valid iff it is true in all models. To prove the relations between this system and our earlier ones, it will be necessary to define a number of operations which were not taken as basic. If θ is an operation then we will use θ^n to indicate the result of applying the operation n times; where $\underline{\theta}$ is the object language symbol for an operation, $\underline{\theta}^n$ will stand for n-concatenations of $\underline{\theta}$. More formally, we define $\theta^1(S)$ to be $\theta(S)$ and $\theta^{n+1}(S)$ to be $\theta(\theta^n(S))$; $\underline{\theta}^1(A)$ will be $\underline{\theta}(A)$ and $\underline{\theta}^{n+1}(A)$ will be $\underline{\theta}(\underline{\theta}^n(A))$. Thus, for example, $\underline{\#}^3\underline{T}$ stands for the formula $\underline{\#\ \#\ \#T}$ and in any interpretation $\langle D, I\rangle$, $I[\underline{\#}^3\underline{T}] = \#^3 I[\underline{T}] = D^3$. In general we will define $\underline{T}^n$ to be $\underline{\#}^n T$ and it is easy to verify that $I[\underline{T}^n]$ will always be D^n. Using this fact we can define an n-formula A to be valid iff $A = \underline{T}^n$ *is valid.*

EXERCISE 8. What is $(P\ \#)^n(S)$?

EXERCISE 9. Show that if S is a set of n-tuples, $\mathcal{R}(S) = R^{n-1}(S)$.

In general, to say of an operation on sets of sequences that it is definable in ET means that there is a string of operators of ET $\theta_1 \ldots \theta_k$ such that the result of applying $\theta_1 \ldots \theta_k$ to a set of sequences always produces the result of the operation in question.

EXERCISE 8. Show that the operator π is definable in ET, where $\pi(S) = \{\langle\sigma(2), \sigma(n), \sigma(3) \ldots \sigma(n-1)\sigma(1)\rangle : \sigma \in S\}$ whenever S is a set of n-tuples.

Two extremely useful operators are F_j and its converse $\reflectbox{F}_j$; F_j is the operation which takes a set of sequences and forms the set of sequences obtained by taking the original sequences and moving their jth element to the front; $\underline{\reflectbox{F}}_j$ is the converse operation which takes a set S of sequences and forms the set of sequences which consist of a member of S with its first element placed after the jth element. We will first illustrate the idea behind the definition in the case of a simple example involving a single sequence. Suppose that we want to define F_4 and let us consider a set S whose only element is the sequence $\langle 1, 2, 3, 4, 5, 6, 7\rangle$; $F_4(S)$ should be the set whose only element is the sequence $\langle 4, 1, 2, 3, 5, 6, 7\rangle$. One way to obtain $F_4(S)$ is first to form $R^2(S)$, which is $\{\langle 3, 4, 5, 6, 7, 1, 2\rangle\}$ and then construct $PR^2(S)$ which is $\{\langle 4, 3, 5, 6, 7, 1, 2\rangle\}$, then $P\mathcal{R}PR^2(S)$ which is $\{\langle 4, 2, 3, 5, 6, 7, 1\rangle\}$ and finally $P\mathcal{R}P\mathcal{R}PR^2(S)$ or $(P\mathcal{R})^2PR^2(S)$ which is the desired set, $\{\langle 4, 1, 2, 3, 5, 6, 7\rangle\}$.

More generally, we can show that the following are true:

$$R^{j-2}(S) = \{\langle\sigma(j-1)\sigma(j), \ldots \sigma(n), \sigma(1), \ldots \sigma(j-2)\rangle : \sigma \in S\}$$

$$PR^{j-2}(S) = \{\langle\sigma(j), \sigma(j-1), \sigma(j+1), \ldots \sigma(n), \sigma(1), \ldots \sigma(j-2)\rangle : \sigma \in S\}$$

$$\text{Я}PR^{j-2}(S) = \{\langle\sigma(j-2), \sigma(j), \sigma(j-1), \sigma(j+1), \ldots \sigma(n), \sigma(1), \ldots \sigma(j-3)\rangle : \sigma \in S\}$$

$$P\text{Я}PR^{j-2}(S) = \{\langle\sigma(j), \sigma(j-2), \sigma(j-1), \sigma(j+1), \ldots \sigma(n), \sigma(1), \ldots \sigma(j-3)\rangle : \sigma \in S\}$$

$$(P\text{Я})^{j-2}PR^{j-2}(S) = \{\langle\sigma(j), \sigma(1) \ldots \sigma(j-1), \sigma(j+1), \ldots \sigma(n)\rangle : \sigma \in S\}.$$

Thus for any set of n-tuples S we can define $F_j(S)$, for $j \leqslant n$ to be $(P\text{Я})^{j-2}PR^{j-2}(S)$; for any n-formula A, $\underline{F_j}(A)$ will be $(\underline{P\text{Я}})^{j-2}\underline{PR}^{j-2}(A)$. A similar argument can be given to show that $\underline{\text{ꟻ}}_j(A)$ should be defined as $\underline{\text{Я}}^{j-1}(\underline{RP})^{j-1}(A)$, where A is an n-formula and $j \leqslant n$.

EXERCISE 9. Show that the operation $\text{Я}^{j-1}(RP)^{j-1}$ produces the desired result.

EXERCISE 10. Using the fact that $P^2(S) = \text{Я}R(S) = R\text{Я}(S) = S$, show that $\text{ꟻ}_j F_j(S) = F_j \text{ꟻ}_j(S) = S$.

If S is any set of n-tuples and $j \leqslant n$, then we can show that

$$EF_j(S) = \{\langle\sigma(1), \ldots \sigma(j-1), \sigma(j+1), \ldots \sigma(n)\rangle : \text{for some } d \text{ in } D, \langle\sigma(1), \ldots \sigma(j-1), d, \sigma(j+1), \ldots \sigma(n)\rangle \in S\}.$$

Thus we can define $\underline{E_j}(A)$ for n-formulas A to be $\underline{E}(A)$ if $j = 1$, $\underline{EF_j}(A)$ if $1 < j \leqslant n$, and $-\underline{T}$ otherwise.

EXERCISE 11. Show that the jth cylindrification operation C_j can be defined as $\text{ꟻ}_j \# EF_j$.

Next we will deal with the problem of defining the identity relations we need. We have taken as our basic formula $\underline{D}$ which in any interpretation is assigned the set of pairs whose first and second elements are the same. We will need in general to have formulas $\underline{D}^n_{i,j}$ which are assigned the set of n-tuples in the domain whose ith and jth elements are identical. We note first that $\underline{\#}^{n-2}\underline{D}$ will be assigned the set of n-tuples whose last two elements are identical, and thus $\underline{\mathfrak{R}}^2\underline{\#}^{n-2}\underline{D}$, will be assigned the set of n-tuples whose first two elements are identical. Thus if we place the first two elements of such a sequence in the ith and jth places respectively we will obtain the sequences we want. Thus, we can define $\underline{D}^n_{i,j}$ for $i < j \leqslant n$, to be $\underline{\exists}_j\underline{\exists}_i\underline{\mathfrak{R}}^2\underline{\#}^{n-2}\underline{D}$; if $j > n$, we let $\underline{D}^n_{i,j}$ be $-\underline{T}$.

We are now ready to prove that the expressive power of ET is at least as great as that of CT.

THEOREM 2. *There is a mapping \$ from formulas of* CT *into formulas of* ET *and from interpretation of* CT *into interpretations of* ET *such that if A is a formula of* CT *and $\langle D, I\rangle$ an interpretation of* CT, *then* $I[A] = \mathrm{UIO}(I^{\$}[\$(A)]$.

Less formally, we have a relation between CT formulas and ET formulas such that for any interpretation of the CT formula there is a corresponding interpretation of the ET formula. The interpretations correspond in the sense that the CT interpretation is just the uniform infinite extension of the ET interpretation.

Proof. We will define \$ by recursion on the order of the formulas and will prove the theorem by induction on the order of formulas.

$\$(F^n_i) = F^n_i$.

$\$(\underline{D}_{i,j}) = \underline{D}^j_{i,j}$ (as defined above).

$\$(\underline{C}_jA) = \underline{C}_j(\$(A))$, where the $\underline{C}_j$ on the right is the ET operator defined above, and where if $\$(A)$ is an n-formula $j \leqslant n$.

$\$(\underline{C}_jA) = \$(A)$ if $\$(A)$ is an n-formula and $j > n$.

$\$(-A) = -\$(A)$.

$\$(A \wedge B) = \$(A) \wedge \$(B)$ if $\$(A)$ and $\$(B)$ are both n-formulas for some n.

$\$(A \wedge B) = \$(A) \wedge \underline{R}^{n-m}\,\underline{\#}^{n-m}\$(B)$, if $\$(A)$ is an n-formula and $\$(B)$ is an m-formula and $n > m$.

$\$(A \wedge B) = \underline{R}^{m-n}\underline{\#}^{m-n}\$(A) \wedge \$(B)$, if $\$(A)$ is an n-formula and $\$(B)$ is an m-formula and $n < m$.

$\$(A \vee B)$ is defined similarly to $\$(A \wedge B)$ with $\vee$ replacing $\wedge$.

$\$(\underline{T}) = \underline{T}$.

To obtain the interpretation $\langle D, I^{\$}\rangle$ from $\langle D, I\rangle$ we need only restrict the assignments to atomic letters to the appropriate finite sequences. $\langle D, I^{\$}\rangle$ is the assignment such that $I^{\$}(F_i^n) = I(F_i^n) \cap D^n$.

We will now prove by induction that $I[A] = \text{UIO}(I^{\$}[S(A)]$. If A is atomic then we know that $I^{\$}(F_i^n) = I(F_i^n) \cap D^n$. Since it is required in the definition of an interpretation for CT that $I(F_i^n)$ be uniform beyond n, we know that $\text{UIO}(I(F_i^n) \cap D^n) = I(F_i^n)$. We also know that $I^{\$}(\$(\underline{D}_{i,j})) = I^{\$}(\underline{D}^j_{i,j})$, which is the set of all j-tuples whose ith and jth members are identical. Applying UIO to that set produces the set of all infinite sequences whose ith and jth members are identical, which is what is required to be shown. The case of $\underline{T}$ is trivial.

Assuming now that our hypothesis holds for formulas of order k, we will show that it holds for formulas of order $k+1$. If we consider a formula $-A$ of order $k+1$, then by our induction hypothesis $I[A] = \text{UIO}(I^{\$}[\$(A)]$. Since the complement of $\text{UIO}(S)$ is UIO of the complement of S, we can infer that

$$\begin{aligned} -\text{UIO}(I^{\$}[\$(A)]) &= \text{UIO}(-I^{\$}[\$(A)]) \\ &= \text{UIO}(I^{\$}[-\$(A)]) \\ &= \text{UIO}(I^{\$}[\$(-A)]). \end{aligned}$$

So by the induction hypothesis, $-I[A] = \text{UIO}(I^{\$}[\$(-A)]) = I[-A]$.

EXERCISE 12. Prove the cases for $A \vee B$ and $A \wedge B$.

The case of formulas $\underline{C}_j A$ where A is an n-formula and $j > n$ is trivial since $I[\underline{C}_j A] = I[A]$, so the remaining interesting case is where $j \leq n$. By the induction hypothesis, if $\underline{C}_j A$ is of order $k+1$, we know that $I[A] = \text{UIO}(I^{\$}[\$(A)])$ and hence a sequence σ is in $I[A]$ iff the sequence $\sigma \upharpoonright n$ consisting of the first n elements of σ is in $I^{\$}[\$(A)]$, since A is an n-formula. Thus by Exercise 11, we know that

UIO($\exists_j \# EF_jI^{\$}[\$(A)]) = C_j(I[A])$, which by the definition of \$ is equivalent to UIO($I^{\$}(\underline{C}_jA)]) = I[\underline{C}_jA]$.

Theorem 2 shows a sense in which the expressive power of ET is at least as great as that of CT. We now prove a theorem in the opposite direction. The main work which must be done in the process of proving this theorem is showing that the permutation operators and E can be suitably represented in CT. Thus we will begin by proving that we can define in CT operators $P_{i,j,n}$ such that if S is a set of sequences uniform from n on, $P_{i,j,n}(S)$ is the set of sequences obtained by taking sequences in S and permuting their ith and jth elements.

LEMMA. *If S is a set of sequences uniform from n on, then*

$$\langle\sigma(1),\ldots\sigma(i),\ldots\sigma(j),\ldots\rangle \in S \text{ iff}$$
$$\langle(1),\ldots\sigma(j),\ldots\sigma(i),\ldots\rangle$$
$$\in C_{n+2}C_{n+1}D_{i,n+2}D_{j,n+1}C_jC_iD_{j,n+2}D_{i,n+1}(S).$$

Proof. The details may be easier to follow if the reader grasps the basic motivation: We begin with a sequence in S uniform from n on, then we identify the i-th and $n+1$st arguments and the jth and $n+2$nd, we use cylindrification to form sequences whose ith and jth places vary arbitrarily, then we identify the ith place with the $n+2$nd and the jth with the $n\pm 1$ and finally we 'erase' the $n+1$st and $n+2$nd places by cylindrification. In effect we have moved the ith and jth places out beyond the relevant part of the sequence and then moved them back into each other's places. Given that S is uniform from n on, we know that

$$\sigma \in D_{j,n+2}D_{i,n+1}(S) \text{ iff } \sigma \in S \text{ and } \sigma(i) = \sigma(n+1) \text{ and } \sigma(j) = \sigma(n+2),$$

so

$$\sigma \in S \text{ iff } \langle\sigma(1),\ldots\sigma(i-1),\ d,\ \sigma(i+1),\ldots\sigma(j-1),\ e,\ \ldots\sigma(n),\ \sigma(i),\ \sigma(j),$$
$$\sigma(n+3)\ldots\rangle \in C_jC_iD_{j,n+2}D_{i,n+1}(S),$$

where d and e are arbitrary elements of the domain. Therefore

$$\sigma \in S \quad \text{iff} \quad \langle \sigma(1), \ldots, \sigma(i-1), \sigma(j), \sigma(i+1), \ldots \sigma(j-1), \sigma(i), \ldots \sigma(n), \sigma(i), \sigma(j) \ldots \rangle \in D_{i,n+2}D_{j,n+1}C_jC_iD_{j,n+2}D_{i,n+1}(S),$$

and so finally we establish

$$\sigma \in S \text{ iff } \langle \sigma(1), \ldots \sigma(i-1), \sigma(j), \ldots \sigma(i-1), \sigma(i), \ldots \sigma(n) \ldots \rangle \in C_{n+2}C_{n+1}D_{i,n+2}, D_{j,n+1}C_jC_iD_{j,n+2}D_{i,n+1}(S).$$

DEFINITION. $\underline{P}_{i,j,n}$ is an abbreviation for

$$\underline{C}_{n+2}\underline{C}_{n+1}\underline{D}_{i,n+2}\underline{D}_{j,n+1}\underline{C}_j\underline{C}_i\underline{D}_{j,n+2}\underline{D}_{i,n+1}.$$

THEOREM 3. *There is a mapping % such that for any formula A of* ET *and any* ET *interpretation $\langle I, D\rangle$, there is a* CT *formula $A^{\%}$ and a* CT *interpretation $\langle i^{\%}, D\rangle$ such that $I^{\%}[A^{\%}] = \text{UIO}(I[A])$.*

Proof. We let $I^{\%}$ be defined as follows: For each atomic formula F_i^n, $I\%(F_i^n) = \text{UIO}(I(F_i^n))$. We define $A^{\%}$ recursively by the following clauses:

$(F_i^n)^{\%}$ is F_i^n
$D^{\%}$ is $D_{1,2}$
$T^{\%}$ is T
$(A \vee B)^{\%}$ is $A^{\%} \vee B^{\%}$, if A and B are both n-formulas, $\perp$ otherwise
$(A \wedge B)^{\%}$ is $A^{\%} \wedge B^{\%}$, if A and B are both n-formulas, $\perp$ otherwise
$(-A)^{\%}$ is $-(A^{\%})$
$(\underline{P}A)^{\%}$ is $\underline{P}_{1,2,n}(A^{\%})$ where A is an n-formula
$(\underline{R}A)^{\%}$ is $\underline{P}_{n-1,n,n} \ldots \underline{P}_{2,3,n}\underline{P}_{1,2,n}(A^{\%})$ where A is an n-formula
$(\underline{\text{Я}}A)^{\%}$ is $\underline{P}_{1,2,n}\underline{P}_{2,3,n} \ldots \underline{P}_{n-1,n,n}(A^{\%})$ where A is an n-formula
$(\underline{\#}A)^{\%}$ is $\underline{P}_{1,2,n+1}\underline{P}_{2,3,n+1} \ldots \underline{P}_{n,n+1,n+1}(A^{\%})$ where A is an n-formula
$(\underline{E}A)^{\%}$ is $\underline{P}_{n-1,n,n} \ldots \underline{P}_{2,3,n}\underline{P}_{1,2,n-1}C(A^{\%})$ where A is an n-formula.

To prove the theorem we now use induction on the order of formulas. The cases for atomic formulas follow immediately from the definitions of % and [ ]. The cases of the sentential connectives are straightforward, so we will concern ourselves only with the more difficult cases. If $\underline{P}A$ is of order $k+1$, then the theorem follows by the lemma concerning the definition of $P_{1,2,n}$ and the induction hypothesis. If we consider a formula of the form $\underline{R}A$, of order $k+1$, then by the induction hypothesis we know that $I^{\%}[A^{\%}] = \mathrm{UIO}(I[A])$, that is $\sigma \in I^{\%}[A^{\%}]$ iff $\langle \sigma(1), \ldots \sigma(n)\rangle \in I[A]$ (assuming A is an n-formula). Since a sequence $\langle \sigma(1), \ldots \sigma(n)\rangle$ is in $I[A]$ iff the sequence $\langle \sigma(2) \ldots \sigma(n), \sigma(1)\rangle$ is in $I[\underline{R}A]$ we need only show that the translation has the corresponding effect on infinite sequences uniform beyond n. We note first that $\sigma \in I^{\%}[\underline{P}_{1,2,n}A^{\%}]$ iff there is a $\sigma \in I^{\%}[A^{\%}]$ which is like σ, except that the elements in the first two places have been switched. Applying $P_{2,3,n}$ forms sequences σ'' which differ from those in $I^{\%}[P_{1,2,n}A^{\%}]$ by permuting the second and third elements. Thus in general, a sequence σ will be a member of

$$I^{\%}[\underline{P}_{n-1,n,n} \cdots \underline{P}_{2,3,n}\underline{P}_{1,2,n}A^{\%}] \text{ iff } \langle \sigma(2), \ldots \sigma(n)\sigma(1), \sigma(n+1) \ldots\rangle$$

is a member of $I^{\%}[A^{\%}]$, which is what was to be shown.

If we consider a formula of the form $\underline{E}A$ where A is an n-formula and $\underline{E}A$ is of order $k+1$, then by the induction hypothesis, an infinite sequence is in $I^{\%}[\underline{C}_1A^{\%}]$ iff for some $d\langle d, \sigma(2), \ldots \sigma(n)\rangle \in I[A]$. Since the rest of the translation of E is the same as the translation of R, by the argument in the above paragraph we can conclude that an infinite sequence will be in $I^{\%}[\underline{P}_{n-1,n,n} \cdots P_{1,2,n}C_1(A^{\%})]$ just in case for some d, $\langle d, \sigma(1), \ldots \sigma(n-1)\rangle \in I[A]$, which is what we needed to prove.

EXERCISE 13. Prove the cases for $\underline{\mathfrak{A}}A$ and $\underline{\#}\,A$.

Having proved Theorem 3 we have clarified the relation between the three systems of quantification theory, cylindrification theory and ET. The reader might wonder whether the translation we have given is the best possible however, for the translation of RA has the property that the length of the translation depends on what type of formula A is. More specifically, the greater the n for which A is an n-formula the longer the translation of R is in RA. Thus, although there is a uniform *method* of translation there is a slightly different translation for RA for each different category of formula A. This is perhaps not surprising on further reflection since the language ET has a finite basis whereas the language CT had infinitely many distinct operations. In fact, it can be shown that our translation is optimal and that the variation in translation reflects an intrinsic difference between the languages.

EXERCISE 14. Show that for any operator θ definable in CT, there is an n such that $\theta(S)$ is uniform beyond n. Using this fact show that no translation of R and Я could translate either of them as single operations in CT.

One of the results we can obtain via the theorems proved in this chapter is that the systems CT and ET have complete axiomatizations and are compact. It is on the basis of these theorems that we can claim that the systems CT and ET are alternative analyses of the connections between syntax and semantics of first order logic. I trust that (if you are not already convinced) further experience in working with these systems will convince the reader that CT and ET are more perspicuous and explicit analyses of the phenomenon underlying the usual formulations of logic.

One way in which the present system is a more explicit analysis of the operations underlying logic is that ordinary formulations of first order logic distinguish only the quantificational portion from the truth functional portion. In the system ET we further distinguish the operations which shorten or lengthen sequences, # and E, and the operations which perform permutations on the sequences, P, R, and Я.

CHAPTER XIV

ANADIC LOGIC

In the last chapter we presented several systems which are essentially equivalent to first order quantification theory with identity. In this chapter we will discuss a natural generalization of those theories which is slightly stronger than standard quantification theory. The viewpoint developed in the last chapter is that logic is the study of operations on sets of sequences and the ways in which those semantic operations can be represented in languages. These operations take the interpretations assigned by a model to the predicate letters and assign sets of sequences to the complex formulas. If one begins from the type of language found in quantification theory where atomic formulas are written in the form $F^n x_1 \ldots x_n$ then it is natural to interpret predicate letters by assigning sets of n-tuples. However, if we take a fresh look at the language which was developed at the end of the last chapter, it is clear that there is no reason to make this restriction. The only remaining trace of the fact that each quantificational predicate letter has a specified number of arguments is in the superscript on predicate letters. Thus in the system to be presented now we will drop the superscripts and also the assumption that predicate letters are assigned sets of n-tuples for some fixed n. Instead, predicate letters will be assigned sets of finite sequences. Thus a possible interpretation of F in a domain $\{0, 1, 2\}$ would be $\{\langle\ \rangle, \langle 0\rangle, \langle 0, 1\rangle, \langle 0, 1, 2\rangle, \langle 0, 0, 0, 0\rangle\}$. Another possible interpretation would be $\{\langle 0\rangle, \langle 0, 0\rangle, \langle 0, 0, 0\rangle \ldots\}$. The reader should be careful to note that the operations in this system are *generalizations* of those in the previous system – they are defined on a wider domain. The importance of this point will be made clearer later when we show that some sets of operations which are interdefinable in the previous system are not in the new one. In order to help the reader remember that the systems differ, we will use **E** as the name of the object language symbol in AL which corresponds to $\underline{E}$ in ET. In general, **A** will be the AL formula corresponding syntactically to the ET formula $\underline{A}$.

The system of anadic logic (AL) will have as vocabulary the following symbols: **T**, −, ∨ , ∧ , **E**, **D**, **P**, **R**, **Я**, #, =, ( ), +, plus an infinite list of predicate letters F_0, F_1, ... We will have no general concept of an n-formula, though we will introduce a related concept shortly. We begin by defining atomic formulas: **T**, **D** and any $\mathbf{F}_n$ are atomic formulas. All atomic formulas are formulas. If **A** and **B** are formulas then **−A**, **(A ∨ B)**, **(A ∧ B)**, **EA**, **PA**, **RA**, **ЯA**, **#A**, and **+A** are formulas. If **A** and **B** are formulas then **A** = **B** is an equation.

Corresponding to the operations symbols of the language we will have operations on sets of finite sequences. Each of the operations except + is the generalization of an operation in ET the only difference between the definitions of these operations and those of the previous chapter being that now S can be any set of finite sequences from the domain.

$$E(S) = \{\langle\sigma(2), \sigma(3), \ldots \sigma(n)\rangle : \sigma \in S\}$$
$$P(S) = \{\langle\sigma(2), \sigma(1), \sigma(3), \ldots \sigma(n)\rangle : \sigma \in S\}$$
$$R(S) = \{\langle\sigma(2), \sigma(3), \ldots \sigma(n), \sigma(1)\rangle : \sigma \in S\}$$
$$Я(S) = \{\langle\sigma(n), \sigma(1), \ldots \sigma(n-1)\rangle : \sigma \in S\}$$
$$\#(S) = \{\langle d, \sigma(1), \ldots \sigma(n)\rangle : d \in D \text{ and } \sigma \in S\}$$
$$+(S) = \langle\sigma(1), \ldots \sigma(n), \sigma(n+1) \ldots \sigma(n+m)\rangle : \langle\sigma(1), \ldots \sigma(n)\rangle \in S\}.$$

The operation + thus forms the set of all finite sequences which are the result of continuing some sequence in S.

An interpretation of the language AL will be an ordered pair $\langle D, I\rangle$ where D is a non-empty set and where I is a function such that $I(\mathbf{F}_n) \subseteq D^f$, where D^f is the set of all finite sequences of objects in D. We can now define $I[\mathbf{A}]$ in the obvious way:

$$I[\mathbf{F}_n] = I(\mathbf{F}_n)$$
$$I[\mathbf{T}] = \mathbf{D}^f$$

$$I[\mathbf{D}] = \{\langle d, d\rangle : d \in D\}$$
$$I[\mathbf{A} \vee \mathbf{B}] = I[\mathbf{A}] \cup I[\mathbf{B}]$$
$$I[\mathbf{A} \wedge \mathbf{B}] = I[\mathbf{A}] \cap I[\mathbf{B}]$$
$$I[-\mathbf{A}] = D^f - I[\mathbf{A}]$$
$$I[\mathbf{EA}] = E(I[\mathbf{A}])$$
$$I[\mathbf{PA}] = P(I[\mathbf{A}])$$
$$I[\mathbf{RA}] = R(I[\mathbf{A}])$$
$$I[Я\mathbf{A}] = Я(I[\mathbf{A}])$$
$$I[\#\mathbf{A}] = \#(I[(\mathbf{A})]$$
$$I[+\mathbf{A}] = +(I[\mathbf{A}]).$$

An equation $\mathbf{A} = \mathbf{B}$ is true in a model $\langle D, I\rangle$ iff $I[\mathbf{A}] = I[\mathbf{B}]$. An equation is valid iff it is true in every model. We can also speak of a formula being valid by an analogy to quantification theory, namely if $I[\mathbf{A}] = D^f$ in every $\langle D, I\rangle$.

The reason that it was necessary to add the + operation to the analogues of the ET operations is that although + is the generalization of an operation which was definable in ET the generalization is not definable from the generalizations of the ET operations. To see this, let us consider the reason why we need +. We often want to form the set of sequences such that, for example, an initial segment of the sequence is assigned to **A** by I and the entire sequence is assigned to **B** by I. For example, if **A** is the set of primes and **B** the relation 'divides evenly', then we might want to form the relation which holds between two numbers just in case the first is a prime and divides the second evenly. In this case **A** would be a 1-formula in ET and **B** would be a 2-formula in ET and we could define the desired set of sequences as $\underline{P(\#(A)) \wedge B}$ since $\underline{I[P(\#(A))]} = \{\langle d_1, d_2\rangle : \langle d_1\rangle \in I[A]\}$ if $\underline{A}$ is a 1-formula in ET. More generally, if we have an n-formula $\underline{A}$ we can find an $n + m$ formula $\underline{C}$ such that the sequences in $I[\underline{C}]$ are exactly those sequences whose first n-elements are a sequence in $I[\underline{A}]$. However, in AL the formulas are not assigned sequences of uniform length nor even of bounded length and thus we cannot always obtain the result we want by applying the operators other than +.

EXERCISE 1. Show that if $\underline{A}$ is an n-formula of ET then for each m there is an $n+m$ formula A' such that for any interpretation $\langle D, I\rangle$, $I[A']$ is the set of sequences of length $n+m$ whose initial n elements are a sequence in $I[A]$.

EXERCISE 2. Show that + is not definable from the other operators in AL.

Our first theorem will show that the system ET is contained in AL. In order to do this we need to verify that various expressions are definable in AL. We will define $\mathbf{V}^{\mathbf{n}}$ analogously to the previous chapter so that $I[\mathbf{V}^{\mathbf{n}}]$ will always be D^n for any $\langle I, D\rangle$. $\mathbf{V}^{\mathbf{0}} = \mathbf{T} \wedge - \#\mathbf{T}$, $\mathbf{V}^{\mathbf{n+1}} = \#\mathbf{V}^{\mathbf{n}}$. We can now define a translation between ET and AL so that there is an AL formula corresponding to each ET formula. We will denote the translation of A as $\mathbf{A}'$; the only problem in the translation is making certain that the formula which results is satisfied only by n-tuples of the appropriate length. Thus we begin by letting $(F_i^n)'$ be $\mathbf{F_2 n_3 i}$. Each of the operations other than negation changes the length of the sequences operated on in the same way as the analogous operation in ET, so we can let $\underline{(A \vee B)'}$ be $\mathbf{A}' \vee \mathbf{B}'$, $\underline{(A \wedge B)'}$ be $\mathbf{A}' \wedge \mathbf{B}'$, $(\underline{PA})'$ be $\mathbf{P(A')}$, $\underline{T'}$ be $\mathbf{T}$, $\underline{D'}$ be $\mathbf{D}$, $\underline{(EA)'}$ be $\mathbf{E(A')}$, $\underline{(RA)'}$ be $(\mathbf{R(A')}$, $\underline{(\text{Я}A)'}$ be $\textbf{Я}\,\mathbf{(A)'}$, and $\underline{(\# A)'}$ be $\#\mathbf{(A')}$. For negation we want to restrict the resulting sequences to the appropriate length so if $\underline{A}$ is an n-formula, then $\underline{(-A)'}$ will be $-\mathbf{(A')} \wedge \mathbf{V}^{\mathbf{n}}$.

THEOREM 1. *For any formula* A *of* ET *there is a formula* A' *of* AL *such that for any interpretation* $\langle D, I\rangle$ *of* ET *there is an interpretation* $\langle D, I'\rangle$ *of* AL *such that* $I[A] = I'[\mathbf{A}']$.

Proof. We let $I'(F_2 i_3 n) = I(F_i^n)$. The theorem can then easily be proved by induction on the complexity of the formula A.

COROLLARY 1. *For any equation* $\underline{A} = \underline{B}$ *of* ET *there is an equation* $\mathbf{A}' = \mathbf{B}'$ *of* AL *such that* $\underline{A} = \underline{B}$ *is valid iff* $\mathbf{A}' = \mathbf{B}'$ *is valid.*

EXERCISE 3. Prove Corollary 1.

COROLLARY 2. *For any formula* $\underline{A}$ *of* ET *there is a formula* $\mathbf{A}'$ *of* AL *such that* $\underline{A}$ *is valid in* ET *iff* $\mathbf{A}'$ *is valid in* AL.

EXERCISE 4. Prove Corollary 2. (Note that $\mathbf{A}'$ is *not* valid whenever $\underline{A}$ is.)

EXERCISE 5. Show that for any formula A of quantification theory there is a formula $\mathbf{A}^*$ of AL such that

(i) A is valid iff $\mathbf{A}^* = \mathbf{V}^0$ is valid

(ii) In any interpretation of AL, $I[\mathbf{A}^*] = \langle\ \rangle$ or $I[\mathbf{A}^*]$ is empty.

Having shown that ET is contained in AL in the sense of Theorem 1, we will now show that the system AL is stronger than ET by showing that AL is not compact.

THEOREM 2. *There are sets of equations* Γ *of* AL *such that* $\Gamma \models \mathbf{B}$ *although for any finite subset* $\Delta \subseteq \Gamma$, $\Delta \models \mathbf{B}$ *does not hold.*

Proof. Let Γ consist of all the sentences $\mathbf{F} \wedge \mathbf{V}^0 = \bot$, $\mathbf{F} \wedge \mathbf{V}^1 = \bot, \ldots$ Then $\mathbf{F} = \bot$ will be a consequence of Γ since the first equation in Γ requires that $\mathbf{F}$ not be assigned the null-sequence, the second that $\mathbf{F}$ is assigned no unit sequence and so on. If, however, we take any finite subset Δ of Γ, then there will be a largest n such that $\mathbf{F} \wedge \mathbf{V}^n = \bot$ is in Δ; Δ can be satisfied by letting $I(\mathbf{F})$ be D^{n+1}, disproving $\Delta \models \mathbf{F} = \bot$.

You should note that the proof that AL is not compact does not make any use of the fact that + is an operation in AL; we will give a different proof of the incompactness of AL shortly in which we make essential use of +. Our next step in exploring the expressive power of AL is to show that there is a particularly simple and elegant method of defining truth for languages based in AL. If you consider the definition of [ ] given on page 152-3 you will note that corresponding to each type of formula there is an equation in which the operator whose role in [ ] is being defined is mentioned on the left and used on the right. That is, we have clauses of the form $I[\mathbf{RA}] = R(I[\mathbf{A}])$. We will now exploit this fact in showing how a particularly simple truth definition can be given in an AL metalanguage for an AL object language.

We will illustrate how to give the definition of truth for an AL language with a finite number of atomic predicates in an AL metalanguage with the same predicates plus some additional vocabulary. Since AL languages do not have individual constants or individual terms the expression of the object language syntax in the metalanguage will be slightly different from the formulation in Chapter IX. In effect we use the method discussed in Chapter IV for expressing what can be expressed using functional terms by means of predicates. Thus the condition of adequacy for expression of the object language syntax in the metalanguage will be that for every formula **A** of the object language there is a complex predicate of the metalanguage $\mathbf{C_A}$ which denotes the formula **A**. In addition to the vocabulary of the object language and sufficient predicates to express the syntax of the object language, we include in the metalanguage one further predicate expression **G**. **G** will be the predicate which we will define to be the satisfaction or denotation relation. We want to give axioms that characterize **G** so that *G* is the relation which holds among an n-tuple of objects just in case the first object is a formula and the remaining $n-1$ objects are a sequence which is in the denotation of the formula. Thus, for example, for each atomic predicate $\mathbf{F_i}$ of the object language we will have an axiom asserting that $\mathbf{E(G \wedge +(C_{`F_i'})) = F_i}$.

Since $\mathbf{C_{`F_i'}}$ is a predicate denoting the atomic formula $\mathbf{F_i}$, $\mathbf{+(C_{`F_i'})}$ will be the set of sequences whose first element is the object language formula $\mathbf{F}_i$. Thus the equation will be true just in case all and only the sequences $\langle d_1, \ldots d_k \rangle$ denoted by **F** are such that $\langle `F_i', d_1, \ldots d_k \rangle$ is denoted by *G*. You should note two features of the definitions that we give. First, it is essential to the definitions that the atomic predicates can be assigned sequences of various lengths. The single predicate letter **G** can relate '$\mathbf{V^1}$' to single objects, '$\mathbf{V^2}$' to pairs of objects and so on. If we put a bound on the length of sequences assigned to **G** then the definition would not be adequate. Further, you should note that in stating the axioms we need to be able to form the set $\mathbf{+C_{`A'}}$ of all finite sequences whose first element is '**A**'. Thus without the expressive power of the + operator we could also not give adequate axioms of the kind to be presented.

The axioms of the denotation and truth theory are:

For each i, $\mathbf{E}(\mathbf{G} \wedge +\mathbf{C}_{`\mathbf{F}_i'}) = \mathbf{F}_i$

$$\mathbf{E}(\mathbf{G} \wedge +\mathbf{C}_{`\mathbf{D}'}) = \mathbf{D}$$

$$\mathbf{E}(\mathbf{G} \wedge +\mathbf{C}_{`\mathbf{T}'}) = \mathbf{T}$$

$$\mathbf{E}(\mathbf{G} \wedge +\mathbf{C}_{`\mathbf{A} \vee \mathbf{B}'}) = \mathbf{E}(\mathbf{G} \wedge +\mathbf{C}_{`\mathbf{A}'}) \vee \mathbf{E}(\mathbf{G} \wedge +\mathbf{C}_{`\mathbf{B}'})$$

$$\mathbf{E}(\mathbf{G} \wedge +\mathbf{C}_{`\mathbf{A} \wedge \mathbf{B}'}) = \mathbf{E}(\mathbf{G} \wedge +\mathbf{C}_{`\mathbf{A}'}) \wedge \mathbf{E}(\mathbf{G} \wedge +\mathbf{C}_{`\mathbf{B}'})$$

$$\mathbf{E}(\mathbf{G} \wedge +\mathbf{C}_{`-\mathbf{A}'}) = -\mathbf{E}(\mathbf{G} \wedge +\mathbf{C}_{`\mathbf{A}'})$$

$$\mathbf{E}(\mathbf{G} \wedge +\mathbf{C}_{`\mathbf{PA}'}) = \mathbf{PE}(\mathbf{G} \wedge +\mathbf{C}_{`\mathbf{A}'})$$

$$\mathbf{E}(\mathbf{G} \wedge +\mathbf{C}_{`\mathbf{RA}'}) = \mathbf{RE}(\mathbf{G} \wedge +\mathbf{C}_{`\mathbf{A}'})$$

$$\mathbf{E}(\mathbf{G} \wedge +\mathbf{C}_{`\text{Я}\mathbf{A}'}) = \text{Я}\mathbf{E}(\mathbf{G} \wedge +\mathbf{C}_{`\mathbf{A}'})$$

$$\mathbf{E}(\mathbf{G} \wedge +\mathbf{C}_{`\mathbf{EA}'}) = \mathbf{EE}(\mathbf{G} \wedge +\mathbf{C}_{`\mathbf{A}'})$$

$$\mathbf{E}(\mathbf{G} \wedge +\mathbf{C}_{`\#\mathbf{A}'}) = \#\mathbf{E}(\mathbf{G} \wedge +\mathbf{C}_{`\mathbf{A}'})$$

$$\mathbf{E}(\mathbf{G} \wedge +\mathbf{C}_{`+\mathbf{A}'}) = +\mathbf{E}(\mathbf{G} \wedge +\mathbf{C}_{`\mathbf{A}'})$$

$$\mathbf{E}(\mathbf{G} \wedge +\mathbf{C}_{`\mathbf{A}=\mathbf{B}'}) = (\mathbf{E}(\mathbf{G} \wedge +\mathbf{C}_{`\mathbf{A}'}) \wedge \mathbf{E}(\mathbf{G} + \mathbf{C}_{`\mathbf{B}'})) \vee$$
$$(\mathbf{E}(\mathbf{G} \wedge +\mathbf{C}_{`-\mathbf{A}'}) \wedge \mathbf{E}(\mathbf{G} \wedge +\mathbf{C}_{`-\mathbf{B}'}).$$

The last axiom is intended to guarantee that $\mathbf{E}(\mathbf{G} \wedge +\mathbf{C}_{`\mathbf{A}=\mathbf{B}'})$ is true whenever $A = B$ is true and the other axioms are to establish for any $A$ in the object language that $\mathbf{E}(\mathbf{G} \wedge +\mathbf{C}_{`\mathbf{A}'}) = A$ is provable.

The following sample should illustrate both the ideas behind the axioms and the method of proof for the theorem. Let us consider the formula $+(\mathbf{RF}_2 \vee -\mathbf{F}_1)$. By the clauses for + we know that

$$\mathbf{E}(\mathbf{G} \wedge +\mathbf{C}_{`+(\mathbf{RF}_2 \vee -\mathbf{F}_1)'}) = +\mathbf{E}(\mathbf{G} \wedge +\mathbf{C}_{`\mathbf{RF}_2 \vee -\mathbf{F}_1'})$$

and so by the axiom for disjunction

$$\mathbf{E}(\mathbf{G} \wedge +\mathbf{C}_{`+(\mathbf{RF}_2 \vee -\mathbf{F}_1)'}) = +(\mathbf{E}(\mathbf{G} \wedge +\mathbf{C}_{`\mathbf{RF}_2'}) \vee \mathbf{E}(\mathbf{G} \wedge +\mathbf{C}_{`-\mathbf{F}_1'})).$$

Using the axioms for R and negation next, we find that

$$\mathbf{E}(\mathbf{G} \wedge +\mathbf{C}_{`+(\mathbf{RF}_2 \vee -\mathbf{F}_1)'}) = +(\mathbf{RE}(\mathbf{G} \wedge +\mathbf{C}_{`\mathbf{F}_2'}) \vee$$
$$-\mathbf{E}(\mathbf{G} \wedge +\mathbf{C}_{`\mathbf{F}_1'}))$$

and so finally by the axioms for F_1 and F_2

$$\mathbf{E}(\mathbf{G} \wedge +\mathbf{C}_{`+(\mathbf{RF}_2 \vee -\mathbf{F})'}) = +(\mathbf{RF}_2 \vee -\mathbf{F}_1).$$

THEOREM 3. *For any object language formula* **A**, **(E(G ∧ +C'A') =** **A** *is provable from the axioms for G.*

EXERCISE 6. Prove Theorem 3. (Use induction on the order of formulas.)

THEOREM 4. *For any object language equation* **A = B**, *the formula* **E(G ∧ +C'A=B')** *is true in an interpretation iff* **A = B** *is.*

Proof. By Theorem 3 and the last axiom we know that

$$\mathbf{E(G \wedge +C_{`A+B'}) = (A \wedge B) \vee (-A \wedge -B)}.$$

If the equation **A = B** is true then the right side will denote D^f and the formula is true. On the other hand, if the formula denotes D^f then it follows that $I[A] = I[B]$.

This method of defining the truth and denotation or satisfaction relations are especially simple and direct and are one of the elegant features of the system of anadic logic. Each axiom gives the relevant clause of the definition for formulas containing an operator using that operator itself in the metalanguage. This is a feature which cannot be copied in any direct way in languages such as quantification theory or CT or ET. This raises, of course, the question of how much stronger than these theories the AL system is. Our last theorem answers this question.

In any interpretation I[**+D**] will be the set of sequences whose first two elements are identical. Consequently, I[**R + D**] will be the set of sequences whose first and last members are identical, and I[**+R + D**] will be the set of sequences whose first element is identical with some later element. I[**R + R + D**] will be the set of sequences whose last element is identical with some other element, and I[**+R + R + D**] will be the set of sequences such that *some* pair of elements are identical. Hence I[**−(+R + R + D)**] will be the set of sequences such that no two elements are identical. We will have frequent occasion to refer to the formula **−(+R + R + D)** so we abbreviate it by **H**.

LEMMA 1. **EH = H** *is true in an interpretation* $\langle D, I\rangle$ *iff* D *is infinite.*

Proof. If D is infinite then for every sequence $\langle d_1, \ldots d_n\rangle$ in $I[H]$

there is a $d \neq d_1, d \neq d_2 \ldots d \neq d_n$ in the domain so that $\langle d, d_1, \ldots d_n \rangle$ will be in $I[H]$ and so $\langle d_1 \ldots d_n \rangle$ will be in $I[EH]$. If, on the other hand the domain contains only n-elements, then all sequences in $I[H]$ will be of length at most n and so any of the maximal length sequences $\langle d_1, \ldots d_n \rangle$ will not be in $I[EH]$.

EXERCISE 7. Give a proof different from the earlier one that AL is not compact.

EXERCISE 8. Show that there is an operation θ definable in AL such that if S is a set of one-element sequences, then $\theta(S)$ is the set of all finite sequences such that at least one member of the sequence is in S.

EXERCISE 9. Show that there is an operation θ definable in AL such that if S is a set of one element sequences, then $\theta(S)$ is the set of all finite sequences such that every element of the sequence is in S.

THEOREM 5. *There is no recursively enumerable set of axioms for the valid equations of* AL.

Proof. We use the fact that the set of formulas of quantification theory which are valid in all finite domains is not recursively enumerable. We know by Theorem 1 of this chapter and Theorem 1 of the previous chapter, that for any formula of quantification theory A we can find a formula A' of AL such that A is true in an interpretation $\langle D, I \rangle$ of quantification theory just in case $A' = V^n$ is true in a corresponding interpretation $\langle D, I' \rangle$ of AL. Thus we know that the set of equations of AL which are valid in all finite domains is not recursively enumerable. In fact, we can be somewhat more specific since we can restrict our attention to closed formulas of quantification and so we can conclude that the set of AL equations of the form $\mathbf{A} = \mathbf{V}^0$ which are valid in all finite domains is not recursively enumerable. Furthermore, we can require $\mathbf{A}$ to be of such a form that for any interpretation either $I[\mathbf{A}]$ is empty or $I[\mathbf{A}] = \{\langle \ \rangle\}$ (by the result of Exercise 5). We will show now that if there were a recursive enumeration of the set of valid equations of AL then there would be a recursive enumeration of the set of equations of the form $\mathbf{A}=\mathbf{V}^0$ which are valid in all finite domains.

From the previous lemma, we know that $I[\mathbf{H}]=I[\mathbf{EH}]$ iff the domain is infinite, and so $I[\mathbf{H}\wedge-\mathbf{EH}]$ will be empty iff the domain is infinite. Since $I[V^0]=\{\langle\ \rangle\}$, we know that $I[\mathbf{V}^0\wedge-\mathbf{A}]$ will be empty if $\mathbf{A}=\mathbf{V}^0$. We consider now the formula $(\mathbf{H}\wedge-\mathbf{EH})\wedge+(\mathbf{V}^0\wedge-\mathbf{A})$. If $I[\mathbf{H}]=I[\mathbf{EH}]$, then the left conjunct is assigned the empty set of sequences, and if $I[\mathbf{A}]=I[\mathbf{V}^0]$, then the right conjunct is assigned the empty set of sequences. Hence if the domain is infinite or $\mathbf{A}=\mathbf{V}^0$ is true in an interpretation, $(\mathbf{H}\wedge-\mathbf{EH})\wedge+(\mathbf{V}^0\wedge-\mathbf{A})=\bot$ will be true in that interpretation.

Suppose that the above equation is true in an interpretation: If $\mathbf{A}=\mathbf{V}^0$ is not true in the interpretation, then $\langle\ \rangle\in I[\mathbf{V}^0\wedge-\mathbf{A}]$ and hence $I[+(\mathbf{V}^0\wedge-\mathbf{A})$ will be the set of *all* sequences from the domain. Therefore given the assumption that the equation is true in the interpretation we are led to the conclusion that $I[\mathbf{H}\wedge-\mathbf{EH}]$ is empty, which implies that the domain is infinite. Thus if the equation is true in an interpretation, either the domain is infinite or $\mathbf{A}=\mathbf{V}^0$ is true in the interpretation. Therefore the equation is valid iff for any interpretation, either the domain is infinite or $\mathbf{A}=\mathbf{V}^0$ is true and so the equation is valid iff $\mathbf{A}=\mathbf{V}^0$ is true in all finite domains.

This theorem shows that the system or anadic logic is considerably stronger than the systems studied in the previous chapter. There remain a number of open questions concerning systems weaker than AL. For example it is not known whether there are complete recursive sets of axioms for the system obtained by omitting + from AL or for the system obtained by adding + to CT or ET.

SELECTED BIBLIOGRAPHY

Chapter I
The ideas behind the definition of a Henkin set were first published by Henkin in 'The completeness of the First Order Functional Calculus', *Journal of Symbolic Logic* **14**, 159–166. Leopold Löwenheim was the first to prove that every satisfiable set of formulas is satisfiable in a denumerable domain; his proof has been translated in *From Frege to Gödel: A Source book in Mathematical Logic* (edited by Jean van Heijenoort), Harvard University Press, Cambridge, Mass., 1967, 660 pp. Thoralf Skolem simplified the methods of proof in his paper 'Über die einige Grundlagenfragen der Mathematik', *Skrifter utgitt av det Norske Videnskaps-Akademi i Oslo*, 1929.

Chapter II
The completeness of first order quantification theory was first proved by Gödel in 1930 by rather different methods; his original paper is translated in *From Frege to Gödel*, as is Emil Post's original proof of the completeness of sentential logic dating from 1921.

Chapter III
Our presentation in this chapter closely follows that of Gentzen's 1934–5 papers which have been translated in the *American Philosophical Quarterly* 1964 and 1965. Discussions of some other related systems and further theorems can be found in Kleene, *Introduction to Metamathematics*, D. van Nostrand Co., Princeton, N.J., 1952, 550 pp.

Chapter IV
The proof of the Löwenheim-Skolem follows essentially a 1920 proof of Skolem's theorem which is translated in *From Frege to Gödel*.

Chapter VI
Gödel's incompleteness theorems were published in 1931; the paper was translated in *From Frege to Gödel*. The non-constructive version of the first theorem is due to Kleene, 'A Symmetric Form of Gödel's Theorem', *Indagationes Mathematics* **12** (1950).

Chapter VII
The method of proof of the incompleteness theorem follows closely unpublished work of Saul Kripke's. The system *Q* is based on a system devised by Raphael Robinson in 1950. Rosser's extension of Gödel's theorem was published in the *Journal of Symbolic Logic* **1** (1936), 87–91. Church's proof of the undecidability of first order logic appeared in the same volume, pp. 40–41, with a 'correction' pp. 101–102.

Chapter VIII
Gödel's proof of the second incompleteness theorem was somewhat less detailed than the proof of the first theorem. A more extensive presentation was given in *Grundlagen der Mathematik* by David Hilbert and Paul Bernays, Springer, Berlin, Vol. 1 1934, 471 pp.; Vol. 2, 1939, 498 pp. The most thorough study is in Solomon Feferman's 'Arithmetization of Metamathematics', *Fundamenta Mathematica* **XLIX** (1960). Lob's theorem appeared in the *Journal of Symbolic Logic* **20** (1955), 115–118.

Chapter IX
The theorems in this chapter originally appeared in a 1933 paper by Tarski which is translated in his *Logic, Semantics and Metamathematics*, Oxford University Press, Oxford, 1956, 471 pp.

Chapter X
Most of the theorems of this chapter are due to Kleene, 'Recursive Predicates and Quantifiers', *Transactions of the American Mathematical Society* **53** (1943), 41–73. Craig's theorem first appeared in the *Journal of Symbolic Logic* **22** (1957) and Mostowski's generalization of the Gödel theorem to non-effective sets of axioms in 'On Definable Sets of Positive Integers', *Fundamenta Mathematica* **34** (1947).

Chapter XI
This interpretation of intuitionistic logic and arithmetic is due to Kleene; a more detailed presentation and some related interpretations are given in his *Introduction to Metamathematics*. Heyting, *Intuitionism* (North-Holland Publ. Co., Amsterdam, 1966, 136 pp.), is the best general introduction to intuitionistic philosophical doctrines and mathematical theories.

Chapter XII
The first systematic study of second order logic as a separate theory was by Hilbert and Ackerman, *Principles of Mathematical Logic*, (originally published in 1928, translated into English in 1950) Chelsea, New York, 172 pp. For a more recent discussion, see Alonzo Church, *Introduction to Mathematical Logic* (Princeton University Press, Princeton, 1956). Systems of branching quantifiers were first studied by Henkin, 'Some Remarks on Infinitely Long Formulas', *Infinitistic Methods*, Pergamon Press, Oxford, 1961, 363 pp.

Chapter XIII
Algebraic methods were introduced into logic by Boole in *The Mathematical Analysis of Logic*, Cambridge, Cambridge, 1897, 82 pp. The elimination of variables was first accomplished by Schonfinkel in a paper translated in *From Frege to Gödel*. The most comprehensive survey of cylindrical algebras (the theory of cylindrification), is in *Cylindric Algebra*, by Henkin, Monk and Tarski (North-Holland, Amsterdam, 1971, 508 pp.). The relations between these systems and standard logic is studied thoroughly in Craig's *Logic in Algebraic Form*, (North-Holland, Amsterdam, 1974, 204 pp). Another equivalent system is presented in Quine's 'Variables Explained Away', *Selected Logic Papers*, (Random House, New York, 1966).

Chapter XIV
The system of anadic logic first appeared in Grandy, 'Anadic Logic and English', *Synthese* (1976).

INDEX OF NAMES

INDEX OF SUBJECTS

INDEX OF SYMBOLS

SYNTHESE LIBRARY

Monographs on Epistemology, Logic, Methodology,
Philosophy of Science, Sociology of Science and of Knowledge, and on the
Mathematical Methods of Social and Behavioral Sciences

1. J. M. Bocheński, *A Precis of Mathematical Logic.* 1959, X + 100 pp.
2. P. L. Guiraud, *Problèmes et méthodes de la statistique linguistique.* 1960, VI + 146 pp.
3. Hans Freudenthal (ed.), *The Concept and the Role of the Model in Mathematics and Natural and Social Sciences, Proceedings of a Colloquium held at Utrecht, The Netherlands, January 1960.* 1961, VI + 194 pp.
4. Evert W. Beth, *Formal Methods. An Introduction to Symbolic Logic and the Study of Effective Operations in Arithmetic and Logic.* 1962, XIV + 170 pp.
5. B. H. Kazemier and D. Vuysje (eds.), *Logic and Language. Studies Dedicated to Professor Rudolf Carnap on the Occasion of His Seventieth Birthday.* 1962, VI + 256 pp.
6. Marx W. Wartofsky (ed.), *Proceedings of the Boston Colloquium for the Philosophy of Science, 1961-1962,* Boston Studies in the Philosophy of Science (ed. by Robert S. Cohen and Marx W. Wartofsky), Volume I. 1973, VIII + 212 pp.
7. A. A. Zinov'ev, *Philosophical Problems of Many-Valued Logic.* 1963, XIV + 155 pp.
8. Georges Gurvitch, *The Spectrum of Social Time.* 1964, XXVI + 152 pp.
9. Paul Lorenzen, *Formal Logic.* 1965, VIII + 123 pp.
10. Robert S. Cohen and Marx W. Wartofsky (eds.), *In Honor of Philipp Frank,* Boston Studies in the Philosophy of Science (ed. by Robert S. Cohen and Marx W. Wartofsky), Volume II. 1965, XXXIV + 475 pp.
11. Evert W. Beth, *Mathematical Thought. An Introduction to the Philosophy of Mathematics.* 1965, XII + 208 pp.
12. Evert W. Beth and Jean Piaget, *Mathematical Epistemology and Psychology.* 1966, XII + 326 pp.
13. Guido Küng, *Ontology and the Logistic Analysis of Language. An Enquiry into the Contemporary Views on Universals.* 1967, XI + 210 pp.
14. Robert S. Cohen and Marx W. Wartofsky (eds.), *Proceedings of the Boston Colloquium for the Philosophy of Science 1964-1966, in Memory of Norwood Russell Hanson,* Boston Studies in the Philosophy of Science (ed. by Robert S. Cohen and Marx W. Wartofsky), Volume III. 1967, XLIX + 489 pp.

15. C. D. Broad, *Induction, Probability, and Causation. Selected Papers.* 1968, XI + 296 pp.
16. Günther Patzig, *Aristotle's Theory of the Syllogism. A Logical-Philosophical Study of Book A of the Prior Analytics.* 1968, XVII + 215 pp.
17. Nicholas Rescher, *Topics in Philosophical Logic.* 1968, XIV + 347 pp.
18. Robert S. Cohen and Marx W. Wartofsky (eds.), *Proceedings of the Boston Colloquium for the Philosophy of Science 1966-1968,* Boston Studies in the Philosophy of Science (ed. by Robert S. Cohen and Marx W. Wartofsky), Volume IV. 1969, VIII + 537 pp.
19. Robert S. Cohen and Marx W. Wartofsky (eds.), *Proceedings of the Boston Colloquium for the Philosophy of Science 1966-1968,* Boston Studies in the Philosophy of Science (ed. by Robert S. Cohen and Marx W. Wartofsky), Volume V. 1969, VIII + 482 pp.
20. J.W. Davis, D. J. Hockney, and W. K. Wilson (eds.), *Philosophical Logic.* 1969, VIII + 277 pp.
21. D. Davidson and J. Hintikka (eds.), *Words and Objections: Essays on the Work of W. V. Quine.* 1969, VIII + 366 pp.
22. Patrick Suppes, *Studies in the Methodology and Foundations of Science. Selected Papers from 1911 to 1969.* 1969, XII + 473 pp.
23. Jaakko Hintikka, *Models for Modalities. Selected Essays.* 1969, IX + 220 pp.
24. Nicholas Rescher *et al.* (eds.), *Essays in Honor of Carl G. Hempel. A Tribute on the Occasion of His Sixty-Fifth Birthday.* 1969, VII + 272 pp.
25. P. V. Tavanec (ed.), *Problems of the Logic of Scientific Knowledge.* 1969, XII + 429 pp.
26. Marshall Swain (ed.), *Induction, Acceptance, and Rational Belief.* 1970, VII + 232 pp.
27. Robert S. Cohen and Raymond J. Seeger (eds.), *Ernst Mach: Physicist and Philosopher,* Boston Studies in the Philosophy of Science (ed. by Robert S. Cohen and Marx W. Wartofsky), Volume VI. 1970, VIII + 295 pp.
28. Jaakko Hintikka and Patrick Suppes, *Information and Inference.* 1970, X + 336 pp.
29. Karel Lambert, *Philosophical Problems in Logic. Some Recent Developments.* 1970, VII + 176 pp.
30. Rolf A. Eberle, *Nominalistic Systems.* 1970, IX + 217 pp.
31. Paul Weingartner and Gerhard Zecha (eds.), *Induction, Physics, and Ethics: Proceedings and Discussions of the 1968 Salzburg Colloquium in the Philosophy of Science.* 1970, X + 382 pp.
32. Evert W. Beth, *Aspects of Modern Logic.* 1970, XI + 176 pp.
33. Risto Hilpinen (ed.), *Deontic Logic: Introductory and Systematic Readings.* 1971, VII + 182 pp.
34. Jean-Louis Krivine, *Introduction to Axiomatic Set Theory.* 1971, VII + 98 pp.
35. Joseph D. Sneed, *The Logical Structure of Mathematical Physics.* 1971, XV + 311 pp.
36. Carl R. Kordig, *The Justification of Scientific Change.* 1971, XIV + 119 pp.
37. Milič Čapek, *Bergson and Modern Physics,* Boston Studies in the Philosophy of Science (ed. by Robert S. Cohen and Marx W. Wartofsky), Volume VII. 1971, XV + 414 pp.

38. Norwood Russell Hanson, *What I Do Not Believe, and Other Essays* (ed. by Stephen Toulmin and Harry Woolf), 1971, XII + 390 pp.
39. Roger C. Buck and Robert S. Cohen (eds.), *PSA 1970. In Memory of Rudolf Carnap*, Boston Studies in the Philosophy of Science (ed. by Robert S. Cohen and Marx W. Wartofsky), Volume VIII. 1971, LXVI + 615 pp. Also available as paperback.
40. Donald Davidson and Gilbert Harman (eds.), *Semantics of Natural Language*. 1972, X + 769 pp. Also available as paperback.
41. Yehoshua Bar-Hillel (ed.), *Pragmatics of Natural Languages*. 1971, VII + 231 pp.
42. Sören Stenlund, *Combinators, λ-Terms and Proof Theory*. 1972, 184 pp.
43. Martin Strauss, *Modern Physics and Its Philosophy. Selected Papers in the Logic, History, and Philosophy of Science*. 1972, X + 297 pp.
44. Mario Bunge, *Method, Model and Matter*. 1973, VII + 196 pp.
45. Mario Bunge, *Philosophy of Physics*. 1973, IX + 248 pp.
46. A. A. Zinov'ev, *Foundations of the Logical Theory of Scientific Knowledge (Complex Logic)*, Boston Studies in the Philosophy of Science (ed. by Robert S. Cohen and Marx W. Wartofsky), Volume IX. Revised and enlarged English edition with an appendix, by G. A. Smirnov, E. A. Sidorenka, A. M. Fedina, and L. A. Bobrova. 1973, XXII + 301 pp. Also available as paperback.
47. Ladislav Tondl, *Scientific Procedures*, Boston Studies in the Philosophy of Science (ed. by Robert S. Cohen and Marx W. Wartofsky), Volume X. 1973, XII + 268 pp. Also available as paperback.
48. Norwood Russell Hanson, *Constellations and Conjectures* (ed. by Willard C. Humphreys, Jr.). 1973, X + 282 pp.
49. K. J. J. Hintikka, J. M. E. Moravcsik, and P. Suppes (eds.), *Approaches to Natural Language. Proceedings of the 1970 Stanford Workshop on Grammar and Semantics*. 1973, VIII + 526 pp. Also available as paperback.
50. Mario Bunge (ed.), *Exact Philosophy – Problems, Tools, and Goals*. 1973, X + 214 pp.
51. Radu J. Bogdan and Ilkka Niiniluoto (eds.), *Logic, Language, and Probability. A Selection of Papers Contributed to Sections IV, VI, and XI of the Fourth International Congress for Logic, Methodology, and Philosophy of Science, Bucharest, September 1971*. 1973, X + 323 pp.
52. Glenn Pearce and Patrick Maynard (eds.), *Conceptual Chance*. 1973, XII + 282 pp.
53. Ilkka Niiniluoto and Raimo Tuomela, *Theoretical Concepts and Hypothetico-Inductive Inference*. 1973, VII + 264 pp.
54. Roland Fraïssé, *Course of Mathematical Logic* – Volume 1: *Relation and Logical Formula*. 1973, XVI + 186 pp. Also available as paperback.
55. Adolf Grünbaum, *Philosophical Problems of Space and Time*. Second, enlarged edition, Boston Studies in the Philosophy of Science (ed. by Robert S. Cohen and Marx W. Wartofsky), Volume XII. 1973, XXIII + 884 pp. Also available as paperback.
56. Patrick Suppes (ed.), *Space, Time, and Geometry*. 1973, XI + 424 pp.
57. Hans Kelsen, *Essays in Legal and Moral Philosophy*, selected and introduced by Ota Weinberger. 1973, XXVIII + 300 pp.
58. R. J. Seeger and Robert S. Cohen (eds.), *Philosophical Foundations of Science. Proceedings of an AAAS Program, 1969*, Boston Studies in the Philosophy of

Science (ed. by Robert S. Cohen and Marx W. Wartofsky), Volume XI. 1974, X + 545 pp. Also available as paperback.

59. Robert S. Cohen and Marx W. Wartofsky (eds.), *Logical and Epistemological Studies in Contemporary Physics*, Boston Studies in the Philosophy of Science (ed. by Robert S. Cohen and Marx W. Wartofsky), Volume XIII. 1973, VIII + 462 pp. Also available as paperback.
60. Robert S. Cohen and Marx W. Wartofsky (eds.), *Methodological and Historical Essays in the Natural and Social Sciences. Proceedings of the Boston Colloquium for the Philosophy of Science, 1969-1972*, Boston Studies in the Philosophy of Science (ed. by Robert S. Cohen and Marx W. Wartofsky), Volume XIV. 1974, VIII + 405 pp. Also available as paperback.
61. Robert S. Cohen, J. J. Stachel and Marx W. Wartofsky (eds.), *For Dirk Struik. Scientific, Historical and Political Essays in Honor of Dirk J. Struik*, Boston Studies in the Philosophy of Science (ed. by Robert S. Cohen and Marx W. Wartofsky), Volume XV. 1974, XXVII + 652 pp. Also available as paperback.
62. Kazimierz Ajdukiewicz, *Pragmatic Logic*, transl. from the Polish by Olgierd Wojtasiewicz. 1974, XV + 460 pp.
63. Sören Stenlund (ed.), *Logical Theory and Semantic Analysis. Essays Dedicated to Stig Kanger on His Fiftieth Birthday*. 1974, V + 217 pp.
64. Kenneth F. Schaffner and Robert S. Cohen (eds.), *Proceedings of the 1972 Biennial Meeting, Philosophy of Science Association*, Boston Studies in the Philosophy of Science (ed. by Robert S. Cohen and Marx W. Wartofsky), Volume XX. 1974, IX + 444 pp. Also available as paperback.
65. Henry E. Kyburg, Jr., *The Logical Foundations of Statistical Inference*. 1974, IX + 421 pp.
66. Marjorie Grene, *The Understanding of Nature: Essays in the Philosophy of Biology*, Boston Studies in the Philosophy of Science (ed. by Robert S. Cohen and Marx W. Wartofsky), Volume XXIII. 1974, XII + 360 pp. Also available as paperback.
67. Jan M. Broekman, *Structuralism: Moscow, Prague, Paris*. 1974, IX + 117 pp.
68. Norman Geschwind, *Selected Papers on Language and the Brain*, Boston Studies in the Philosophy of Science (ed. by Robert S. Cohen and Marx W. Wartofsky), Volume XVI. 1974, XII + 549 pp. Also available as paperback.
69. Roland Fraïssé, *Course of Mathematical Logic* – Volume 2: *Model Theory*. 1974, XIX + 192 pp.
70. Andrzej Grzegorczyk, *An Outline of Mathematical Logic. Fundamental Results and Notions Explained with All Details*. 1974, X + 596 pp.
71. Franz von Kutschera, *Philosophy of Language*. 1975, VII + 305 pp.
72. Juha Manninen and Raimo Tuomela (eds.), *Essays on Explanation and Understanding. Studies in the Foundations of Humanities and Social Sciences*. 1976, VII + 440 pp.
73. Jaakko Hintikka (ed.), *Rudolf Carnap, Logical Empiricist. Materials and Perspectives*. 1975, LXVIII + 400 pp.
74. Milič Čapek (ed.), *The Concepts of Space and Time. Their Structure and Their Development*, Boston Studies in the Philosophy of Science (ed. by Robert S. Cohen and Marx W. Wartofsky), Volume XXII. 1976, LVI + 570 pp. Also available as paperback.

75. Jaakko Hintikka and Unto Remes, *The Method of Analysis. Its Geometrical Origin and Its General Significance*, Boston Studies in the Philosophy of Science (ed. by Robert S. Cohen and Marx W. Wartofsky), Volume XXV. 1974, XVIII + 144 pp. Also available as paperback.
76. John Emery Murdoch and Edith Dudley Sylla, *The Cultural Context of Medieval Learning. Proceedings of the First International Colloquium on Philosophy, Science, and Theology in the Middle Ages – September 1973*, Boston Studies in the Philosophy of Science (ed. by Robert S. Cohen and Marx W. Wartofsky), Volume XXVI. 1975, X + 566 pp. Also available as paperback.
77. Stefan Amsterdamski, *Between Experience and Metaphysics. Philosophical Problems of the Evolution of Science*, Boston Studies in the Philosophy of Science (ed. by Robert S. Cohen and Marx W. Wartofsky), Volume XXXV. 1975, XVIII + 193 pp. Also available as paperback.
78. Patrick Suppes (ed.), *Logic and Probability in Quantum Mechanics*. 1976, XV + 541 pp.
79. H. von Helmholtz, *Epistemological Writings*. (A New Selection Based upon the 1921 Volume edited by Paul Hertz and Moritz Schlick, Newly Translated and Edited by R. S. Cohen and Y. Elkana), Boston Studies in the Philosophy of Science, Volume XXXVII. 1977 (forthcoming).
80. Joseph Agassi, *Science in Flux*, Boston Studies in the Philosophy of Science (ed. by Robert S. Cohen and Marx W. Wartofsky), Volume XXVIII. 1975, XXVI + 553 pp. Also available as paperback.
81. Sandra G. Harding (ed.), *Can Theories Be Refuted? Essays on the Duhem-Quine Thesis*. 1976, XXI + 318 pp. Also available as paperback.
82. Stefan Nowak, *Methodology of Sociological Research: General Problems*. 1977, XVIII + 504 pp. (forthcoming).
83. Jean Piaget, Jean-Blaise Grize, Alina Szeminska, and Vinh Bang, *Epistemology and Psychology of Functions*. 1977 (forthcoming).
84. Marjorie Grene and Everett Mendelsohn (eds.), *Topics in the Philosophy of Biology*, Boston Studies in the Philosophy of Science (ed. by Robert S. Cohen and Marx W. Wartofsky), Volume XXVII. 1976, XIII + 454 pp. Also available as paperback.
85. E. Fischbein, *The Intuitive Sources of Probabilistic Thinking in Children*. 1975, XIII + 204 pp.
86. Ernest W. Adams, *The Logic of Conditionals. An Application of Probability to Deductive Logic*. 1975, XIII + 156 pp.
87. Marian Przełęcki and Ryszard Wójcicki (eds.), *Twenty-Five Years of Logical Methodology in Poland*. 1977, VIII + 803 pp. (forthcoming).
88. J. Topolski, *The Methodology of History*. 1976, X + 673 pp.
89. A. Kasher (ed.), *Language in Focus: Foundations, Methods and Systems. Essays Dedicated to Yehoshua Bar-Hillel*, Boston Studies in the Philosophy of Science (ed. by Robert S. Cohen and Marx W. Wartofsky), Volume XLIII. 1976, XXVIII + 679 pp. Also available as paperback.
90. Jaakko Hintikka, *The Intentions of Intentionality and Other New Models for Modalities*. 1975, XVIII + 262 pp. Also available as paperback.
91. Wolfgang Stegmüller, *Collected Papers on Epistemology, Philosophy of Science and History of Philosophy*, 2 Volumes, 1977 (forthcoming).

92. Dov M. Gabbay, *Investigations in Modal and Tense Logics with Applications to Problems in Philosophy and Linguistics.* 1976, XI + 306 pp.
93. Radu J. Bogdan, *Local Induction.* 1976, XIV + 340 pp.
94. Stefan Nowak, *Understanding and Prediction: Essays in the Methodology of Social and Behavioral Theories.* 1976, XIX + 482 pp.
95. Peter Mittelstaedt, *Philosophical Problems of Modern Physics,* Boston Studies in the Philosophy of Science (ed. by Robert S. Cohen and Marx W. Wartofsky), Volume XVIII. 1976, X + 211 pp. Also available as paperback.
96. Gerald Holton and William Blanpied (eds.), *Science and Its Public: The Changing Relationship,* Boston Studies in the Philosophy of Science (ed. by Robert S. Cohen and Marx W. Wartofsky), Volume XXXIII. 1976, XXV + 289 pp. Also available as paperback.
97. Myles Brand and Douglas Walton (eds.), *Action Theory. Proceedings of the Winnipeg Conference on Human Action, Held at Winnipeg, Manitoba, Canada, 9-11 May 1975.* 1976, VI + 345 pp.
98. Risto Hilpinen, *Knowledge and Rational Belief.* 1978 (forthcoming).
99. R. S. Cohen, P. K. Feyerabend, and M. W. Wartofsky (eds.), *Essays in Memory of Imre Lakatos,* Boston Studies in the Philosophy of Science (ed. by Robert S. Cohen and Marx W. Wartofsky), Volume XXXIX. 1976, XI + 762 pp. Also available as paperback.
100. R. S. Cohen and J. Stachel (eds.), *Leon Rosenfeld, Selected Papers.* Boston Studies in the Philosophy of Science (ed. by Robert S. Cohen and Marx W. Wartofsky), Volume XXI. 1977 (forthcoming).
101. R. S. Cohen, C. A. Hooker, A. C. Michalos, and J. W. van Evra (eds.), *PSA 1974: Proceedings of the 1974 Biennial Meeting of the Philosophy of Science Association,* Boston Studies in the Philosophy of Science (ed. by Robert S. Cohen and Marx W. Wartofsky), Volume XXXII. 1976, XIII + 734 pp. Also available as paperback.
102. Yehuda Fried and Joseph Agassi, *Paranoia: A Study in Diagnosis,* Boston Studies in the Philosophy of Science (ed. by Robert S. Cohen and Marx W. Wartofsky), Volume L. 1976, XV + 212 pp. Also available as paperback.
103. Marian Przełęcki, Klemens Szaniawski, and Ryszard Wójcicki (eds.), *Formal Methods in the Methodology of Empirical Sciences.* 1976, 455 pp.
104. John M. Vickers, *Belief and Probability.* 1976, VIII + 202 pp.
105. Kurt H. Wolff, *Surrender and Catch: Experience and Inquiry Today,* Boston Studies in the Philosophy of Science (ed. by Robert S. Cohen and Marx W. Wartofsky), Volume LI. 1976, XII + 410 pp. Also available as paperback.
106. Karel Kosík, *Dialectics of the Concrete,* Boston Studies in the Philosophy of Science (ed. by Robert S. Cohen and Marx W. Wartofsky), Volume LII. 1976, VIII + 158 pp. Also available as paperback.
107. Nelson Goodman, *The Structure of Appearance,* Boston Studies in the Philosophy of Science (ed. by Robert S. Cohen and Marx W. Wartofsky), Volume LIII. 1977 (forthcoming).
108. Jerzy Giedymin (ed.), *Kazimierz Ajdukiewicz: Scientific World-Perspective and Other Essays, 1931–1963.* 1977 (forthcoming).
109. Robert L. Causey, *Unity of Science.* 1977, VIII+185 pp.
110. Richard Grandy, *Advanced Logic for Applications.* 1977 (forthcoming).

111. Robert P. McArthur, *Tense Logic*. 1976, VII + 84 pp.
112. Lars Lindahl, *Position and Change: A Study in Law and Logic*. 1977, IX + 299 pp.
113. Raimo Tuomela, *Dispositions*. 1977 (forthcoming).
114. Herbert A. Simon, *Models of Discovery and Other Topics in the Methods of Science*, Boston Studies in the Philosophy of Science (ed. by Robert S. Cohen and Marx W. Wartofsky), Volume LIV. 1977 (forthcoming).
115. Roger D. Rosenkrantz, *Inference, Method and Decision*. 1977 (forthcoming).
116. Raimo Tuomela, *Human Action and Its Explanation. A Study on the Philosophical Foundations of Psychology*. 1977 (forthcoming).
117. Morris Lazerowitz, *The Language of Philosophy*, Boston Studies in the Philosophy of Science (ed. by Robert S. Cohen and Marx W. Wartofsky), Volume LV. 1977 (forthcoming).
118. Tran Duc Thao, *Origins of Language and Consciousness*, Boston Studies in the Philosophy of Science (ed. by Robert S. Cohen and Marx. W. Wartofsky), Volume LVI. 1977 (forthcoming).
119. Jerzy Pelc, *Polish Semiotic Studies, 1894–1969*. 1977 (forthcoming).
120. Ingmar Pörn, *Action Theory and Social Science. Some Formal Models*. 1977 (forthcoming).
121. Joseph Margolis, *Persons and Minds*, Boston Studies in the Philosophy of Science (ed. by Robert S. Cohen and Marx W. Wartofsky), Volume LVII. 1977 (forthcoming).

SYNTHESE HISTORICAL LIBRARY

Texts and Studies
in the History of Logic and Philosophy

Editors:

1. M. T. Beonio-Brocchieri Fumagalli, *The Logic of Abelard.* Translated from the Italian. 1969, IX + 101 pp.
2. Gottfried Wilhelm Leibniz, *Philosophical Papers and Letters.* A selection translated and edited, with an introduction, by Leroy E. Loemker. 1969, XII + 736 pp.
3. Ernst Mally, *Logische Schriften,* ed. by Karl Wolf and Paul Weingartner. 1971, X + 340 pp.
4. Lewis White Beck (ed.), *Proceedings of the Third International Kant Congress.* 1972, XI + 718 pp.
5. Bernard Bolzano, *Theory of Science,* ed. by Jan Berg. 1973, XV + 398 pp.
6. J. M. E. Moravcsik (ed.), *Patterns in Plato's Thought. Papers Arising Out of the 1971 West Coast Greek Philosophy Conference.* 1973, VIII + 212 pp.
7. Nabil Shehaby, *The Propositional Logic of Avicenna: A Translation from al-Shifā: al-Qiyās,* with Introduction, Commentary and Glossary. 1973, XIII + 296 pp.
8. Desmond Paul Henry, *Commentary on De Grammatico: The Historical-Logical Dimensions of a Dialogue of St. Anselm's.* 1974, IX + 345 pp.
9. John Corcoran, *Ancient Logic and Its Modern Interpretations.* 1974, X + 208 pp.
10. E. M. Barth, *The Logic of the Articles in Traditional Philosophy.* 1974, XXVII + 533 pp.
11. Jaakko Hintikka, *Knowledge and the Known. Historical Perspectives in Epistemology.* 1974, XII + 243 pp.
12. E. J. Ashworth, *Language and Logic in the Post-Medieval Period.* 1974, XIII + 304 pp.
13. Aristotle, *The Nicomachean Ethics.* Translated with Commentaries and Glossary by Hypocrates G. Apostle. 1975, XXI + 372 pp.
14. R. M. Dancy, *Sense and Contradiction: A Study in Aristotle.* 1975, XII + 184 pp.
15. Wilbur Richard Knorr, *The Evolution of the Euclidean Elements. A Study of the Theory of Incommensurable Magnitudes and Its Significance for Early Greek Geometry.* 1975, IX + 374 pp.
16. Augustine, *De Dialectica.* Translated with Introduction and Notes by B. Darrell Jackson. 1975, XI + 151 pp.